Exploring for the Small Business and Home Office

by Louis Columbus

Other PROMPT® Books by Louis Columbus

Administrator's Guide to E-Commerce
Exploring the World of SCSI

Exploring LANs for the Small Business and Home Office

by Louis Columbus

PROMPT®　PUBLICATIONS

©2001 by Sams Technical Publishing

PROMPT® Publications is an imprint of Sams Technical Publishing, 5436 W. 78th St., Indianapolis, IN 46268.

All rights reserved. No part of this book shall be reproduced, stored in a retrieval system, or transmitted by any means, electronic, mechanical, photocopying, recording, or otherwise, without written permission from the publisher. No patent liability is assumed with respect to the use of the information contained herein. While every precaution has been taken in the preparation of this book, the author, the publisher or seller assumes no responsibility for errors or omissions. Neither is any liability assumed for damages resulting from the use of information contained herein.

International Standard Book Number: 0-7906-1247-X
Library of Congress Catalog Card Number: 2001093161

Acquisitions Editor: Deborah Abshier
Senior Editor: Kim Heusel
Editor: Ruth Frick, BooksCraft Inc.
Interior Design: Debbie Berman
Typesetting: Debbie Berman
Indexing: Kim Heusel
Proofreader: Jan Zunkel
Cover Design: Christy Pierce
Graphics Conversion: Christy Pierce
Illustrations: Courtesy the author

Trademark Acknowledgments:
All product illustrations, product names and logos are trademarks of their respective manufacturers. All terms in this book that are known or suspected to be trademarks or services have been appropriately capitalized. PROMPT® Publications and Sams Technical Publishing cannot attest to the accuracy of this information. Use of an illustration, term or logo in this book should not be regarded as affecting the validity of any trademark or service mark.

PRINTED IN THE UNITED STATES OF AMERICA
9 8 7 6 5 4 3 2 1

Contents

1 Fundamentals of LANs for Small Business and Home Offices 1

Market Dynamics Driving the Growth of Small Business and Home Networking 2
Comparing Home Networking Technologies 4
    Ethernet Networks 4
    Power-Line Networks 6
        Key Power-Line Vendors and Technologies 7
        Adaptive Networks 7
    Phone-Line Networks 9
        Avio Digital 11
        Key Phone-Line Vendors and Technologies 12
    Wireless Technologies 13
        IEEE 802.11 14
        HomeRF 16
        Key Wireless Home Networking Vendors 18
Working with Wireless Networking 22
    Benefits of Wireless LANs 22
    Wireless Technologies Overview 23
    Wireless LAN Technology Options 24
        Spread Spectrum 24
        Narrowband Technology 25
        Frequency-Hopping Spread-Spectrum Technology 25
        Direct-Sequence Spread-Spectrum Technology 25

   Infrared Technology 26
   Microcells and Roaming 26
   Wireless LAN 27
 Summary 32

2   Choosing a LAN Cabling Solution for Your Home or Office 35

 Case Study: Wireless Networking Quickly
 Adopted by Ice Cream Maker *35*
  Using Copper Wiring 37
   Copper UTP Wiring: What Is It? 38
  Using 8-Pin Modular (RJ-45) 41
  Working with Video Cables 42
  Infrared 43
   The Advantages of Infrared 43
   Benefits of Infrared 43
   The IrDA Standard 44
   Connecting with IR 44
   Summary of Infrared 44
   A Commitment to Infrared Technology 45

3   Introduction to TCP/IP and Networking in Windows 2000 47

 Introduction 47
 A Brief History of TCP/IP 48
 How TCP/IP Fits into Microsoft's Networking Strategy 49
 TCP/IP Basic Concepts 49
  Hardware Independence 50
  Standardized Addressing 50

  Open Standards 51
  Application Protocols 52
 Introducing the OSI Model 52
  Physical Layer 54
  Data Link Layer 55
  Network Layer 55
  Transport Layer 56
   Session Layer 56
  Presentation Layer 57
  Application Layer 57
 Understanding How TCP/IP Enables Communication 57
  IP Addressing 58
   IP Address Classes 59
   What Is Subnet Masking? 59
  IP Routing 61
  Getting to Know IP Multiplexing 62
Integrating Microsoft TCP/IP
into a Network 63
TCP/IP Connectivity Utilities 64
 Finger Service 64
 FTP 64
 TFTP 68
 Telnet 68
 RCP 71
 Rexec 72
 Rsh 72
 LPR 73
The Dynamic Host
Configuration Protocol 73
 What Is DHCP? 74
 Why DHCP Was Created 76
 Understanding How DHCP Works 76
  Installing DHCP Client Services 78
 Using DNS in Windows 2000 80
  How DNS Naming Resolution Happens 81
  The Role of the HOSTS File 82
Summary 82

4 Introducing Home and Small Business 85

LAN Connectivity Basics 85
Definitions and Standards 86
Access and Collisions 88
Problem Determination 90
Frame Formats 91
    Ethernet II or DIX 92
    IEEE 802.3 and 802.2 93
    Snap 94
Conclusion 94

5 Managing Serial Devices in a Networked Environment 97

Management via Serial Communications 97
    Nonnetworked Techniques 97
    Networked Techniques 98
Device Server Technology 99
The Benefits of Networked Management 100

6 How Home and Small Business LANs Use Addressing 103

Subnetting 106
Public and Private IP Addresses 110
IP Overview 111
    Layers 1 and 2 112
    Layer 3 112
        Internet Protocol 112
        Address Resolution Protocol 114
    Layer 4 114
        Transmission Control Protocol 115

  User Datagram Protocol 115
  Internet Control and Messaging Protocol 115
 Layers 5, 6, and 7 116
  TCP Applications 116
  UDP Applications 116

7 Exploring the Fundamentals of Transmission Media for Home and Small Businesses 119

Introduction 119
Exploring Types of
Transmission Frequencies 120
Discovering Transmission
Media Characteristics 121
 Cost 121
 Installation Requirements 122
 Bandwidth 122
 Band Usage (Baseband or Broadband) 123
  Multiplexing 124
 Attenuation 124
 Electromagnetic Interference 124
Comparing Cable Media 125
 Coaxial Cable 125
  Types of Coaxial Cable 126
  Coaxial Characteristics 127
  Coax and Fire Code Classifications 130
 Twisted-Pair Cable 131
  Unshielded Twisted-Pair Cable 133
 Fiber-Optic Cable 136
  Fiber-Optic Characteristics 137
Learning about Wireless Media 138
 Reasons for Using Wireless Networks 139
 Wireless Communications
 with LANs 140

   Infrared Transmission 142
   Laser Transmission 142
   Narrowband Radio Transmission 143
   Spread-Spectrum Radio Transmission 143
   Microwave Transmission 144
  Summary 145

8 Administrator's Guide to Managing a Small Business LAN 147

Approaching a Problem 148
  Troubleshooting Hints 149
  Diagnostic Tools 151
   Testing Basic Connectivity 152
   Diagnostic Tools Summary 154
  Troubleshooting with the ifconfig Command 154
  Network Hardware Problems 155
  Checking Routing 155
Summary 157

9 How LANs Use Protocols and Layering 159

What Is an Internetwork? 159
History of Internetworking 160
Internetworking Challenges 161
Open Systems Interconnection
Reference Model 161
  Characteristics of the OSI Layers 162
OSI Model and Communication
between Systems 164
Interaction between OSI Model Layers 164
  OSI-Layer Services 165

OSI Model Layers and Information Exchange 165
  Information Exchange Process 166
  Physical Layer 168
  Data Link Layer 168
  Network Layer 170
  Transport Layer 170
  Session Layer 171
  Presentation Layer 171
  Application Layer 172
Information Formats 172
ISO Hierarchy of Networks 175
Connection-Oriented and Connectionless Network Services 175
  Connection-Oriented Services 176
    Disadvantages 177
    Advantages 177
  Connectionless Services 177
Internetwork Addressing 178
  Data Link–Layer Addresses 178
  MAC Addresses 178
  Network-Layer Addresses 181
  Hierarchical Versus Flat Address Space 181
    Address Assignments 182
  Addresses Versus Names 182
Flow-Control Basics 184
Error-Checking Basics 185
Multiplexing Basics 185

10 INTERNETWORKING CONCEPTS FOR SMALL BUSINESS AND HOME LANS 187

A New Phase for Home Networking 187
  Home Networking Primer 188
What Is a Gateway and Why Do I Need One? 189
  Communication 190
  Productivity 191

Security and Home Monitoring 192
Entertainment 192
Choosing the Best Network
Gateway for Your Home 193
3Com HomeConnect Home Networking Solutions 193
3Com Ethernet, Phone-Line, and Wireless
Home Networking Products 194
3Com HomeConnect Cable and DSL Modems 194
New 3Com HomeConnect Gateway Offerings 195
Summary 197
Endnotes 197

11 Knowing the Fundamentals of TCP/IP for Running Your Network 199

Getting TCP/IP Up and Running 200
Hands-on Tutorial: Getting TCP/IP Working in Windows 2000
Professional 201
Installing TCP/IP 201
Configuring TCP/IP for DHCP Support 203
Understanding Domain Name Services System 205
Understanding Differences between
DHCP and DNS 206
Installing SNMP 208
Architecture of the Windows
Internet Naming Services 211
Resolving Names on Microsoft Networks 212
Naming on a WINS Network 214
Managing WINS Servers 214
Adding WINS Server to WINS Server Manager 215
Monitoring WINS 215
Viewing WINS Server Details 216
Configuring Static Mappings 217
Managing the WINS Database 219
Replicating the WINS Database 219

xii

    Maintaining the WINS Database 220
        Backing Up the WINS Database 221
        Restoring the WINS Database 222
    Managing LMHOSTS Files 223
        Format of LMHOSTS Files 223
        LMHOSTS Keywords 224
        Enabling Clients to Use LMHOSTS Files 225
        Guidelines for Establishing LMHOSTS
        Name Resolution 225
    Summary 226

12 Evaluating Wireless Networks for Your Home or Office 227

Forecasts by e-Market Dynamics 228
    Subscriber Forecast 229
    Revenue Forecast 229
    Major Forecast Assumptions 233
The Competitive Advantage
of Going Wireless 234
    IEEE 802.11 and 802.11b Technology 236
        802.11 Operating Modes 237
        The 802.11 Physical Layer 238
        802.11b Enhancements to the Physical
        Address (PHY) Layer 240
        The 802.11 Data Link Layer 241
Current Standards in Wireless Networking 247
    What's New in Wireless LANs:
    The IEEE 802.11b Standard 247
Considerations for Choosing a WLAN 248
    Ease of Setup 248
    Ease of Management 249
    Range and Throughput 250
    Mobility 250
    Power Management 251

xiii

Safety	251
Security	252
Cost	252
Summary	253
Endnote	253

Appendix A
Sun Microsystem's View of the Connected Home 255

Appendix B
Using the Internet Connection Wizard in Microsoft Small Business Server 275

Glossary 309

Index 315

Introduction

Beginning with the birth of the Internet, networking has steadily grown from its first incarnation as a means of communicating throughout a single enterprise to what we have today—a means of sharing information between distant locations around the world. Networking is the basis of the Internet, and, with the rapid developments being made in broadband, wireless, and gigabit technologies, networking is going to be a dominant part of everyone's life within the next 20 years. It will change how the world communicates, conducts business, coordinates activities, and ultimately builds relationships. A major part of this change is just starting to happen today, with the advent of wireless technologies and their impact on the home just now being seen through the pioneering efforts of Linksys and other manufacturers building kits that enable homes to have wireless networks within them.

Driving the adoption of networking in both small businesses and home offices is the fact that the world is becoming an increasingly interdependent place, especially when it comes to marketing, commerce, and fulfillment of operations. Clearly small businesses are benefiting from networking today, even if they haven't yet installed a network of their own. With websites recommending the best small businesses and the impact of email on everyday life, word of mouth is now being accentuated with e-mails that voice approval for a new bakery or store. Because word of mouth drives revenue for many small businesses, networking is streamlining that aspect of how businesses get recommended. A second factor driving networking growth in small businesses is their need to share information across geographic distances quickly. Recent studies completed by International Data Corporation (IDC) indicate that the small businesses with the highest growth rate are those that can create a unified messaging and information-sharing vision that, in turn, guides the rollout of each additional office. The study also concludes that the majority of high-growth companies actively utilize broadband and wireless technologies to be more efficient at serving their customers. With the growth of companies as well as their efficiency being linked with their ability to be more responsive through networking, the future has become clear. Networking for small businesses and offices is transforming the nature of work itself. This book concentrates on the necessary technologies for getting your small business, home office, or even your home networked between personal computers and the outside world.

EXPLORING LOCAL AREA NETWORKING

At its most fundamental, a network can be two PCs connected with each other for the purpose of sharing of information and resources. The most common reason small businesses get a network up and running is to share files and printers. As simple as a shared drive and as complex as an application that supports hundreds of users, networking is taking the concept of instant information and bringing it to small businesses and home offices. Regarding home offices specifically, today there are more telecommuters than ever before, according to IDC, making networking between corporate workplace and home office a necessity. The seamlessness between Internet access at home and work gives workers more flexibility than ever to pursue both family and work activities on their own schedules. One manager I know takes a day a week at home to catch up on e-mails and to finish projects. The high-speed access of the Internet in conjunction with his own network at home gives him the privacy to get more work done in less time.

Others with home offices—some making the decision to forego corporate life to start their own small companies out of their homes—quickly find that networking and speed of response is critical for growing new businesses. Networking is all about bringing speed of response and accuracy of communications to customers.

THIS BOOK'S MISSION

Use this book as a guide for getting your network up and running in your small business or home office. Instead of relating the history behind networking and its progression, this book aims to give you the kind of direction you need to get your PCs talking to each other. You will also learn how to make fax machines, printers, plotters, and even Internet-enabled TVs talk to one another. In short, anything with an Internet Protocol (or IP address for short) can be part of a network. You will also get a chance to see how the experts in this area go to school to get certifications so that they can keep companies from 100 to 100,000 employees humming. If you want to pursue certification, information is in the appendices of this book for you. After reading through the hands-on examples and steps for getting businesses online, you should have a strong foundation of knowledge for growing in this area.

First, the goal of giving you pragmatic advice and guidance in setting up a network reverberates through each chapter. Along the way to this knowledge, the journey needs to be fun too, so there are humorous examples throughout this book of what can go wrong with a network as well as some of the more humorous aspects of networks as they get installed into a company or even a home. So, keep your sense of humor with you. As you work on your network, it will keep your goal of getting a networking up and running much easier to accomplish.

This book will have been a success if at the end you get the knowledge you came here to get and had fun in the process.

CHAPTER-BY-CHAPTER TOUR

Taking the steps from the fundamentals to the advanced aspects of networking your business and your home, this book starts with the concepts of media and cabling, as well as various technology discussions regarding wireless and networking. With the foundation set, each chapter progresses through the fundamentals of building a local area network, or LAN, followed by directions on how to share resources throughout a network.

Chapter 1 discusses the fundamentals of networking, including how wireless technologies work, a discussion of the different types of media and cabling, and a description of how home networking technologies have quickly evolved into both wireless and more traditional wired approaches to enable communication between multiple PCs in a home and the Internet. This chapter also includes examples of network applications in your home and brief coverage of the evolving role of service providers who provide access to services via the Internet.

Chapter 2 focuses on the task of choosing the best possible cabling solution for your business or home. Getting a good handle on concepts is always easier when case studies are used for communicating concepts and real-world scenarios. In this chapter, a case study provides insights into how builders are including broadband connectivity in the homes they are constructing today. In one development in Irvine, California, for example, there is a closet specifically designed for a server! This is evidence of how pervasive home networking is, even being designed into new homes and developments. Chapter 2 ends with a comparison of radio, wireless, microwave, and infrared (IR) for connecting devices together in a network.

Chapter 3 provides a foundation for creating a network in your business or home, starting with the concept of creating a network in your home with hands-on examples. The chapter assumes that many of you will be using the Microsoft Windows 2000 Professional operating system to handle the task of getting a system up and running on a network.

Chapter 4 focuses on the LAN connectivity basics and essentials of small business and home office networking. This chapter discusses how networking works from the protocol level and includes descriptions of how to detect errors in transmissions (thankfully, many tools are available for handling this task today, so you don't have to do it on your own!). The chapter ends with a brief description of how to troubleshoot errors in communication on your LAN.

Chapter 5 gives you a glimpse into how networks communicate with each other using serial interface protocols. The very beginnings of networking in homes was started by Apple with their LocalTalk protocol, which is still serially based where bits of data are sent one after another on a network. One of the most important aspects of creating a network within your business or home is defining the addressing approach that will be used. Chapter 6 focuses on how to use the LAN hardware filter packets using addresses, defining how a physical address works. It also describes how you can streamline the performance of your network using addressing.

Chapter 7 walks you through the decisions of how to define the connections between your PCs, which are the essential elements of a network. This chapter gets to the hands-on aspects of networking and describes the options for cabling types and the essentials of how Network Interface Cards (NICs) work. You'll also find this chapter full of hands-on advice for installing and configuring NICs. While many companies do provide installation manuals, this chapter provides lessons learned from completing installations throughout a multioperating-system network.

The administrator is the person who keeps a network humming, and chapter 8 gives you a glimpse into the world of administrators by taking you through the techniques they use for solving many of the configuration issues they encounter. This chapter discusses how administrators cope with and solve problems of cable length limitations and how companies are increasingly using wireless technologies to make it easier to communicate with companies via PCs and laptops. This book will provide a glimpse into how administrators have actually contributed to the widespread use of wireless networking via their continued adoption of this technology. Wireless has significant potential in home networking as well.

Chapter 9 provides a tutorial on how LANs use protocols and layering; while this sounds much more technical that it is, the intent is to provide, in layman's terms, a brief description of how networks speak with one another and use commands to make information sharing happen.

Using the concept of sharing resources—Internet access, files, printers, and even interactive games—chapter 10 focuses on the internetworking concepts that you can use in planning your LAN. This chapter begins with a description of how internetworking works, followed by the concepts of Universal Service in networks and a description of how to use internetworking with legacy data to make your network even more valuable through the sharing of information. There is also a section specifically dedicated to routers and their role in networking. Just as people of various nations speak widely varying languages, the same holds true for networks. Chapter 11 provides guidance in the area of the most heavily used protocol, TCP/IP. Don't be intimidated by this term—you will know it well by the end of the book. It's really nothing more than an approach to handling the task of making two or more PCs communicate with each other more effectively. One of the greatest benefits of basing your network on TCP/IP is that it is much more scalable than any of the other networking technologies, and it is also more secure. This chapter guides you through configuring TCP/IP within the Windows 2000 Professional environment for handling local area networking in a home or office.

The impact of wireless communications in networking is a central theme throughout this book, and chapter 12 provides the finishing touch on the subject. The fact that today even servers are supporting wireless communication is covered in this chapter, in addition to considerations for designing a wireless-based network. There is also an overview of how to create a network using wireless technologies, including a case study from Ericsson on its latest developments in wireless products.

IS THIS BOOK FOR YOU?

Developed specifically to meet the needs of people building networks in their small businesses, this book focuses on hands-on examples of putting a network together. However, the book also targets networking enthusiasts who want to get a handle on how to create networks in their homes and share PCs between rooms. For example, I discuss at length the exciting development of wireless technology, its use in networking, and how it has made it possible to have PCs throughout a home or office networked together without cabling. I also point out the leading providers of networking products, with recommendations based on my own personal experience of creating a LAN in my own home.

KEY TERMS AND CONCEPTS

Networking has its own language, and the acronyms in this book will be kept to a minimum to ensure ease of reading. When getting started on a new subject, it's always a good idea to get a handle on the most commonly used terms and concepts, so take a few minutes to look at the glossary at the back of the book to familiarize yourself with these terms and their definitions—they will be used frequently.

CHAPTER 1

Fundamentals of LANs for Small Business and Home Offices

Let's say that you have a small business or a home office that could benefit from having its computers linked so that they could communicate with each other, and could have capabilities similar to the networks used by large companies, referred to as local area networks or LANs. How can you accomplish this? You can get a network up and running in your business or home using a series of wireless devices that communicate through radio frequencies, or you can use devices that are linked together using wires, like many traditional networks that are created today. Taking the building blocks of a network and explaining each one of them makes the actual hands-on work much easier to accomplish and understand. The goal of this chapter is to provide an overview of just what's involved in a network and how networks became what they are today. This chapter starts with a brief overview of the market dynamics behind the growth of networking, and follows with an explanation of the key concepts that make a LAN actually work.

Throughout this chapter are also short vignettes and case studies of businesses just like yours that were able to get their networks up and running quickly; these will show you the practical side of building a network. The focus is on the key foundational concepts of building a network, anchored in actual experiences.

MARKET DYNAMICS DRIVING THE GROWTH OF SMALL BUSINESS AND HOME NETWORKING

First, to answer the question of just why so many small businesses and homes are getting LANs installed, we must look at the accelerating pace of change in how organizations have been communicating over the Internet, starting in the early 1990s and progressing to today. The cost of having broadband or high-speed access to the Internet at our homes has plummeted in the last 10 years. Today it's possible for a home to have the same, if not better, network access speeds compared to an office or even a small corporation of 100 or more people. That's because broadband technologies are becoming increasingly available for anyone willing to create his or her own network.

An amazing trend is developing with the growth of LANs in homes and small businesses: The number of kits sold, which provide everything needed in a single box, is actually being overtaken in sales by individual components. Companies such as Linksys now offer a router that includes a firewall for under $200. This price point brings the functionality of router technology within the reach of many homes and small businesses. Figure 1.1 provides an image of the Linksys BEFSR41, one of the leading routers available on the market today.

The thing that's really driving the overall adoption of networking is the growing realization that sharing Internet access, printers, file servers, and even multimedia servers, can be accomplished more economically now than ever before. Growth factors include the following:

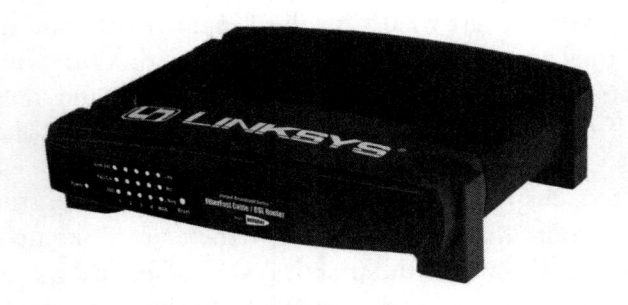

FIGURE 1.1 Linksys BEFSR41

High demand for Internet access—Today, broadband connectivity to the Internet is more pervasive than ever before, and it is growing rapidly throughout many areas of the world. The term *broadband* refers to Internet access speeds at or above 10 megabytes per second (Mbps), which is significantly faster than the dial-up modems used by most people for getting access to the Internet and other online services. There is going to be at least a threefold jump in home networking throughout the next decade as more and more families decide to share a single high-speed Internet access connection throughout their homes.

Sharing files—The very first networks in companies were built specifically for sharing files. These first networks were called peer-to-peer networks or, as they are known today, P2P. One of the most common uses of all first networks was the sharing of files. In your home, how is this different? The file sharing is done the same way; the difference is merely in the sizes and uses of the files shared. So, in the world of home networking, this first, most dominant use of networking means that you can move the family album, game files—in short, anything—over the network.

Printer sharing–Being able to spread the costs of printing over many different users has long been an approach businesses have taken to getting the most out of their investments in peripherals.

CD sharing–Being able to share the contents of both music and data CDs throughout a household is made possible through the use of networks and servers. Increasingly, game CDs are used for creating a multiuser environment where all members of a family can have access to a game together.

Multimedia file sharing (entertainment)—More than any other factor, gaming and the use of networked, highly graphical games in homes is making networking technologies more desirable than ever before. Networked versions of *Doom*, for example, were the favorites of developers, where they could battle each other in a virtual world over their lunch periods. Networked versions of games are growing faster than anticipated, thanks in large part to the rapidly dropping costs of networking products.

COMPARING HOME NETWORKING TECHNOLOGIES

E-Market Dynamics believes that the market is shaping up to support four technologies: Ethernet, phone line, power line, and wireless (detailed in Table 1.1).

ETHERNET NETWORKS

Home Ethernet technology requires the same type of cabling that runs through the walls and ceilings of most U.S. business offices. It has a reputation for security, reliability, and speed. However, Ethernet has also earned a reputation for being difficult to install, often requiring holes to be drilled in walls and floors and/or wires to be run under carpets. The Ethernet network is also difficult to manage. The arrival of Ethernet in-a-box kits in 1999 changed this picture somewhat. These kits offer low-cost Ethernet adapter cards, an inexpensive nonmanaged hub to direct network traffic, and basic networking software designed to make installation and use simpler; however, many households will still need wiring.

Transmission speeds of available Ethernet home networking kits range from 10 Mbps to 100 Mbps, currently the fastest of any home networking products. Ethernet home networks require Category 3 or Category 5 unshielded twisted pair (UTP) copper wire cabling between connected devices.

Ethernet offers two main advantages over other home networking technologies: adherence to the proven, well-supported IEEE 802.3 standard (making the technology outlook secure) and reliable data transmission speeds ranging from 10 Mbps to 100 Mbps.

Nevertheless, significant disadvantages remain with Ethernet home networks. They are often expensive to install due to the need for special wiring and a hub; but, in many cases, they are still cheaper than new alternatives. The task of wiring and setting up the network can lead to potential headaches. Few homes possess a network-ready environment (unlike offices with cubicles and a dedicated staff); installing the wiring can also lead to aesthetics problems.

TABLE 1.1 U.S. Home Networking Technology Specifications

Technology	Ethernet	Phone Line	Power Line	HomeRF	802.11	Bluetooth
Current data rate (Mbps)	10/100	10	350 Kbps	2	11	0.72
Future data rate (Mbps)	NA	100	10	10	NA	10 or 20
Availability	Now	Now	Now	Now	Now	2H00
Proponent		HomePNA	HomePlug Powerline Alliance			
Pros	Secure, reliable, and fast	Many backers	Easy to use, flexible access	Mobile, easy to use	Fast data transmissions	Will be used in multiple devices
	One industry standard	Endorsed by top OEMs	Available; power lines ubiquitous	Voice applications	Mobile, easy to use	Lower cost
		Easy integration into other silicon	Lower cost	Lower cost		
Cons	Requires new wiring	Phone jacks not ubiquitous	Multiple standards, competing technologies	Limited range	More expensive	Currently not robust enough to be a full home network
			HomePlug must choose standard quickly	Slower data throughput		Product not yet available
			Slow speeds, subject to interference			

Source: e-Market Dynamics, 2000

Several vendors, including D-Link Systems, Intel, Kingston, Linksys, NETGEAR, Protec, 3Com, and UMAX, have introduced Ethernet home networking kits. Examples of this technology include the Fast Ethernet 10/100 Network in a Box from Linksys and the DFE-910 Network in a Box from D-Link (both of which cost less than $100; each includes two 32-bit Peripheral Component Interconnect (PCI) LAN cards and a 5-port hub). There is also the 3Com Homeconnect Home Network 10-Mbps USB Ethernet kit, priced at $190.

POWER-LINE NETWORKS

Power-line networks use existing electrical wiring within walls to transmit data throughout a house, offering the convenience of plugging nodes directly into any home electrical outlet. However, this convenience comes at a cost; no power-line networks currently shipping are able to transmit data faster than 350 kilobits per second (Kbps), far slower than other home networking technologies. The technology does hold some potential for evolution: Intellon, Enikia, and ITRAN have recently announced fast (10- 12-Mbps) power-line modem chips. While power-line technology may not evolve into the preferred data networking solution, the wide availability of electrical outlets could make it a simple and inexpensive application for other types of networks (such as home automation).

Power-line networks offer several advantages. With an electrical outlet positioned every eight feet (on average), access is very flexible. The physical labor involved in installing power-line networks is minimal—one need only plug the appliance/adapter into the electrical outlet. Vendors are working to create packages that support an easy installation and maintenance process. Finally, this is a relatively low-cost technology, with a cost per node ranging from $40 to $150. Even with faster throughput speeds, the HomePlug Powerline Alliance expects pricing to be in a similar range; more significantly, the price will be comparable to Home Phoneline Networking Alliance (HomePNA) pricing.

Nevertheless, significant disadvantages exist for power-line home networking technology. As previously mentioned, currently available technology is very slow, with data throughput of 350 Kbps at most. Moreover, power-line-based networking is subject to many types of interference that could cause signal loss. For example, sudden voltage dips or surges may result from a lightning strike or, more commonly, a high-energy household resource (such as an air conditioner, refrigerator, or hair dryer). These power dips and surges could potentially "fry" devices not plugged into a surge protector and/or cause data degradation. Security—always an issue—is also problematic for power-line technology. Data on an unencrypted household network may be accessible to neighbors sharing the same electrical transformer. Finally, and perhaps most important, the power-line home network industry is beset by a lack of standards. As they stand, competing power-line network technologies are incompatible with each other and, with the exception of Ethernet, with other networking solutions. The lack of an industry standard has not drawn wide support for the technology, especially in comparison with other home networking technologies.

Some work has been started to address these disadvantages. Vendors have begun to discuss and demonstrate faster technology. The HomePlug Powerline Alliance was formed in April 2000 to set a standard for the technology and to promote it within the marketplace. The alliance plans to have a specification in place by the end of the year. According to e-Market Dynamics, this will not be an easy task, because the process of choosing one technology over several might lead to severe complications; these issues could potentially be avoided if vendors are willing to make concessions and/or share technologies. This may also delay an already rigorous time line.

Key Power-Line Vendors and Technologies

The following are some of the key vendors and technologies that have been introduced in the power-line networking market:

ADAPTIVE NETWORKS

Adaptive Networks develops and markets chipsets, modules, and end-user products that incorporate spread-spectrum technology at speeds up to 115 Kbps.

ENIKIA. Enikia has demonstrated the Information Appliance Network (IAN). The IAN is designed to connect all types of electronic devices (white goods, consumer electronics, utilities, telecommunications, and the Internet) at speeds of up to 10 Mbps through power lines. Enikia's 10-Mbps Powerline Ethernet Transceiver chipset adapts standard, off-the-shelf Ethernet controllers to the home's power-line network, enabling up to 256 devices to be connected in homes of up to 5,000 square feet in size. Network interface adapters are then attached to electronic devices through USB or parallel ports so that the devices can communicate with each other as well as connect to the Internet. The network is secure because the signal is encrypted before transmission. In addition, the signal attenuates quickly, so it will not leave the home and is thus potentially more secure than other solutions. The company expects to distribute products through communication service providers and Original Equipment Manufacturers (OEMs).

INARI. Formerly Intelogis, Inari Inc.'s PassPort plug-in adapters offer up to 350 Kbps of data throughput over AC power lines. A lack of commercial success recently forced Inari to shift its focus from the development of consumer products to the development of silicon and other components for OEMs. It expects to develop power-line chipsets rated at 4 Mbps in 2002 and to have a 10-Mbps chipset in wide distribution by 2002. Inari is focused on selling its chipsets to modem, gateway, and Network Interface Card (NIC) vendors and to consumer electronics OEMs.

ITRAN COMMUNICATIONS. Based in Tel Aviv, ITRAN has developed power-line modems—the ITM1 and ITM10—that are capable of speeds of up to 2.5 Mbps and 12 Mbps, respectively, over residential power lines. The physical and data-link layers are specially designed to work with Ethernet application layers. Thus, the replacement of dedicated cable LANs with existing electrical wiring is transparent to the end user. The ITM10 modulates Ethernet applications onto the power-line carrier, enabling file transfer, print sharing, and other networking applications without changing existing software. The ITM10 thus constitutes a power-line extension to Ethernet or HomePNA, enabling the seamless transmission of compatible applications over residential power lines. Evaluation kits are now available.

INTELLON. Intellon has developed the Power Packet Home Networking Technology with speeds of up to 11 Mbps. The company is ramping up their current products with major product introductions planned for May, 2002. It is based on proprietary spread-spectrum carrier technology; the power-line integrated circuits (ICs) are optimized for communications over AC or DC power-line networks.

MEDIA FUSION. Media Fusion is developing proprietary technology that enables data to ride along the electrical wires. In this technology, signals travel in the magnetic field surrounding the wires (versus other technologies that have attempted to send signals through electrical wires along with the electric current, which generally corrupts the signal). In theory, this technique will avoid signal loss and distortion. Media Fusion claims that data can be passed along electrical wires at up to 2.5 gigabits per second (Gbps), a nearly unimaginable speed. Therefore, it is not surprising that this technology remains unproven. Nevertheless, Media Fusion rolled out a last-mile solution in the second half of 2000, using electrical power grids to deliver low-cost, very-high-speed access to Internet, telephone, and television services. The consumer technology would consist of modemlike interface controllers and device controllers that could be built into devices such as personal computers (PCs), televisions, and phones. Media Fusion estimates that the average cost per household for its adapters will be under $60.

PHONE-LINE NETWORKS

The two technologies in use today for phone-line networks are HomePNA (by far the more pervasive) and Avio Digital.

HomePNAPhone-line home networking technology is based on HomePNA 2.0, the de facto industry standard. The technology is robust, with a rating of up to 10 Mbps and the ability to support up to 500 feet of phone wire between devices connected to RJ-11 jacks. HomePNA technology is relatively easy to use: The user must install software and set up the network through both the software settings and the physical connection to the phone jack. Finally, security is excellent because each home has at least one unique phone circuit (phone number) from the phone company's central office. The downside is that homes with multiple phone lines are, in most cases, limited to confining the network to one phone line.

The current HomePNA 2.0 specification was introduced in late 1999 by HomePNA. This revised spec is backward compatible with the 1.0 spec (introduced in 1998). While it does have a throughput rating of 10 Mbps, products show actual throughput in the 7- to 8-Mbps range. HomePNA 2.0 may not be sufficiently fast for users who engage heavily in streaming video applications, but it is sufficient for most home networking applications. A HomePNA 2.0 network can support up to 25 PCs, peripherals, and other network devices.

The HomePNA 2.0 specification is based on technology from Broadcom and Lucent. Broadcom supplies iLine10 chipsets, each of which code a packet of data so that the receiver can properly decode it. Like Ethernet, HomePNA uses a packet-based architecture, which is interoperable with HomeRF and Ethernet networks. Total POTS (short for *plain old telephone service*) bandwidth is segregated into channels for each type of traffic—power, analog voice, and digital information (audio, data, and video)—by filtering technology called *frequency division multiplexing*. HomePNA 2.0 provides quality of service (QoS) that enables streaming audio, video, and telephony over the same wires used for Internet access.

HomePNA networking technology offers several advantages. First and foremost, the alliance is composed of 115 members, including PC and networking companies. This has enabled multiple companies to develop solutions that should be compatible with other HomePNA-based solutions. As consumer awareness increases, a single, well-known standard for phoneline home networking can help alleviate confusion, which should enable more sales. Second, HomePNA solutions are relatively inexpensive, with several solutions priced under $100, enabling PC OEMs to bundle the solutions with PC sales. Third, the elegance of the solution gives credence to the mantra of "no new wires," making installation and network management relatively simple. Finally, data throughput, while rated at 10 Mbps, actually hovers in the 7- to 8-Mbps range. This might be considered a disadvantage, but, because the typical home networks have been running serially based protocols that do not deliver the performance necessary for larger data transfers and more efficient use of Internet bandwidth, HomePNA in the 7- and 8-Mbps range is actually presenting a very robust data throughput solution.

Nevertheless, this technology has some disadvantages, one of which is the limited number of phone jacks available in most homes. Also, the presence of multiple phone lines in a single household presents a challenge because the solution is limited to a single phone line. And while the standard calls for high data throughput rates, the actual product has emerged with much slower speeds. The technology could also lose momentum when it attempts to move beyond the United States; while U.S. households tend to have multiple phone jacks, households in other countries are often limited to one or two phone jacks.

Avio Digital

While HomePNA 2.0 has become the de facto phone-line standard, it is important to note that at least one company has developed technology not based on that standard: Avio Digital. The company is expected to release its proprietary phone-line networking solution in 2000 and be in widespread distribution by 2002. It will license its MediaWire chipsets (at about $10 when purchased in quantity), which provide up to 88 Mbps bandwidth for home networks over Category 3 wires and roughly double that over Category 5 wiring. The chipsets offer ranges of 33 meters between devices using standard Category 3 telephone wiring, 100 meters with Category 5 wiring, or 400 meters with coaxial cable. MediaWire home networks will be capable of supporting up to 100 devices of various types located throughout the home, over a total cable length of up to 10,000 meters. MediaWire implements a distributed architecture that enables network applications to be controlled from *smart* devices (those devices containing processors). Because it is not packet based, devices such as speakers or appliances connecting to the network can be inexpensive *dumb* devices. It supports both analog and digital devices. MediaWire's synchronous architecture allows all types of media to be distributed without loss of quality.

One of the first products to incorporate MediaWire chips is Scientific Atlanta's (SA) Explorer 2000, a digital television set-top box designed to offer two-way communications with interactive applications, Internet support, video on demand, and home shopping. SA distributes the set-top box through 14 cable service providers in North America, including Time Warner and TCI. Using Avio Digital's MediaWire, the Explorer 2000 acts as an interface for connected home theater components, PC networks, home security devices, and consumer electronics via telephone wiring.

Avio Digital's MediaWire technology is very compelling. However, in order to significantly penetrate the home networking market, Avio Digital must align itself with companies that will bundle and promote the technology, including PC OEMs and other prominent home networking players. With much of the industry rallying around the HomePNA standard, this task will be very difficult to achieve. Nevertheless, the technology should not be discounted completely. If HomePNA does not take off in any measurable form, the industry could move toward Avio Digital. Alternatively, the company could form a substantial partnership with vendors that could provide them with a channel through which to bring their products to the mainstream.

Key Phone-Line Vendors and Technologies

The following sections describe the products of some of the key vendors addressing this space with HomePNA-based products:

Diamond Multimedia/S3. Diamond Multimedia (acquired by S3) offers the HomeFree Phoneline 10-Mbps Desktop Pack (PCI). The kit includes two PCI cards and associated software at a list price of $130. The kit is also bundled with select Compaq Presario desktop models.

D-Link Systems. D-Link Systems offers the DHN-910 10-Mb Phoneline Network. The kit, which lists for $119, includes two standard PCI adapters with phone-line cords, software, and RJ-11 jacks.

Intel. Intel has branded its HomePNA technology under the moniker of AnyPoint Home Network. The vendor offers parallel, PCI, and external USB versions of its AnyPoint home phone-line solution. The PCI and USB versions are rated at 10 Mbps, while the parallel port and another USB version are rated at 1 Mbps. Because each kit is used for only one PC, at least two kits must be purchased at $89 each.

LINKSYS. Linksys offers the HomeLink Phoneline Network in a Box. The package includes two PCI cards with RJ-45 jacks, which offer the dual capability of being used in either a HomePNA network or 10BaseT Ethernet network. This package is generally priced under $100.

NETGEAR. This former Nortel subsidiary offers the Home Phone Line PCI Adapter (PA 301) as well as a USB version, both of which are rated at 10 Mbps. The list price is $95 for either version.

3COM. 3Com offers consumers the HomeConnect 10-Mbps PCI Phoneline kit, which includes two PCI adapters rated at 10 Mbps and HomeClick networking software. It lists for $159. Dell also offers the solution as a bundled option with its PCs.

WIRELESS TECHNOLOGIES

Wireless home networking technologies are based on three radio frequency (RF) ranges:

- **902 to 928 megahertz (MHz).** This frequency band is utilized by many cordless phones. Due to a large number of devices using the band, interference is common. While an occasional crackle or pop is acceptable during a phone conversation, this type of interference can cause errors in data transmission.

- **2.400 to 2.483 gigahertz (GHz).** In 1985, the FCC set aside this swath of radio real estate for unlicensed use in industrial, science, and medical (ISM) applications in the United States. This band is used by the majority of wireless networking products. Because this band is used by multiple white goods (such as newer cordless phones and microwave ovens), interference can also be problematic.

- **5.725 to 5.875 GHz.** This band is relatively underused; only RadioLAN currently offers a home networking kit in this range. As the 2.4-GHz band becomes crowded, more wireless vendors will likely introduce products in the 5-GHz band, which offers more available bandwidth and enables greater speed and range than do 2.4-GHz networks. Due to the need to firmly establish standards and the higher costs for the chip technology involved, this band has for the most part been avoided by wireless networking vendors.

RF wireless transmission is based on spread-spectrum technology, a modulation technique that spreads data transmissions across the entire available frequency band in a prearranged scheme. This type of modulation is aimed toward making the signal resistant to noise, interference, and snooping. Spread spectrum preserves signal integrity by detecting collisions and automatically resending any lost data. The technology also permits many users to share a frequency band with minimal interference from other users and other 2.4-GHz devices (such as microwave ovens). Also, most vendors supply PC cards with Wireless Equivalent Privacy (WEP) encryption, either a 40-bit encryption key or a more sophisticated 128-bit key. Two types of spread-spectrum modulation have been specified by the 802.11 standard: direct-sequence spread spectrum (DSSS) and frequency-hopping spread spectrum (FHSS). Later in this chapter, these will be covered in more detail.

IEEE 802.11

The 802.11 standard specifies one media access controller (MAC) protocol and three physical layers: direct sequence at 1 or 2 Mbps, frequency hopping at 1 Mbps, and diffuse infrared. The MAC offers two types of operation: a polling mode or a distributed mode Carrier Sense Multiple Access/Collision Detect (CSMA/CD). There are also options in terms of MAC-level retransmissions and power management. Each vendor may also implement proprietary extensions. All in all, the 802.11 standard is very complex and, considering its varying implementations, it is somewhat difficult to conceive of the 802.11 as a single standard.

The 802.11 standard utilizes DSSS modulation. In data transmission, each bit to be transmitted is encoded with a redundant pattern called a chip; the encoded bits are spread across the entire available frequency band. The encoding ensures that the transmission is known only to the sending and receiving stations, making it difficult for an intruder to intercept and decipher the encoded wireless data. The redundant pattern also makes it possible to recover data without the need to retransmit if one or more bits are damaged or lost during transmission. Roaming is possible only between access points on the same channel, making DSSS less robust for roaming than FHSS. Interoperability is currently supported among only those devices using the same transmission technology.

While seemingly wasteful of bandwidth, DSSS copes well with weak signals: Data can often be extracted from a background of interference and noise without having to be retransmitted, making actual throughput more robust; DSSS can reject noise from a microwave oven with relative ease.

When the IEEE 802.11 standards body came together in September 1997 to establish the 802.11 standard, disagreements within the committee allowed the standard to be issued with three incompatible physical layers. Several wireless vendors insisted on specifications in the standard that made it closer to their own products, and the resulting standard is bloated with features. Furthermore, several vendors have added proprietary extensions to the 802.11 standard.

Initially this complexity slowed development and increased costs, but with standardization in mid-1999 on 802.11b (the 2.4-GHz band with DSSS at 11 Mbps) and growing competition and volume, costs are starting to fall.

IEEE working groups currently have two extensions to the 802.11 standard under development, both of which modify the direct-sequence physical layer but use the standard 802.11 MAC:

- **802.11a.** This extension provides for data rates from 6 to 54 Mbps in the unlicensed 5-GHz frequency band, using orthogonal frequency division multiplexing (OFDM), a spread-spectrum technology developed by Wi-LAN Inc. This technology will probably be ready for market in two years.

- **802.11b.** This extension upgrades data transfers to 11 Mbps in the 2.4-GHz frequency band using DSSS. Networks based on 802.11b are becoming increasingly available from vendors such as Apple, Cisco/ Aironet, Lucent, Cabletron, 3Com, Symbol Technologies, Compaq, and Farallon. List prices of access points range from $299 to $1,299. PC adapter cards are priced from $99 to $295; higher-priced adapters generally include stronger encryption technology.

The Wireless Ethernet Compatibility Alliance (WECA) certifies the interoperability of wireless LANs with the Wi-Fi standard. This certification will be key going forward for wireless home networking technologies because it certifies that differing vendor solutions will work together, having been tested for interoperability by a third-party testing group.

Several advantages exist for 802.11b wireless home networking technologies. It is a robust solution, with maximum data throughput speeds of up to 11 Mbps versus 2 Mbps for HomeRF. This higher throughput rate will prove to be more practical for higher-bandwidth applications such as video. Similar to the advantages discussed earlier for wireless technologies in general, 802.11b provides flexibility and mobility in terms of ease of implementation and use. Finally, 802.11b technology has the advantage of carryover into the corporate space, which will also give it some appeal in the home.

Nevertheless, the technology has at least one notable disadvantage: it can be costly, with a single node costing as much at $300. A less serious disadvantage is that its restricted range can be problematic; in addition, walls can block its signals.

HomeRF

The HomeRF standard utilizes FHSS, in which a transmitting and receiving station are synchronized to hop from channel to channel in a predetermined pseudorandom sequence. This prearranged hop sequence is known only to the transmitting and receiving stations. IEEE 802.11 specifies 79 channels and 78 different hop sequences; if one channel is jammed or noisy, the data is simply retransmitted when the transmitter hops to a clear channel. As a result of its constant frequency shifting, FHSS is less susceptible to interference and is difficult for an intruder to intercept, making it very secure. Because interference is minimal with FHSS, multiple hopping sequences can typically be assigned in the same physical area, allowing more users to share the available bandwidth. This also allows users to roam between access points on different channels.

The HomeRF specification is implemented through an interoperability protocol termed the Shared Wireless Access Protocol (SWAP), which defines a common interface supporting wireless data and voice networking among PCs, peripherals, PC-enhanced cordless phones, and handheld devices. The specification also provides for six PCS-quality voice channels for telephones and a short-range, low-power mode for digital devices such as PDAs. A single connection point can support both voice services via time division multiple access (TDMA) and data services via Carrier Sense Multiple Access/Collision Avoidance (CSMA/CA). HomeRF operates in the 2.45-GHz Interfrequency Switching Mode (ISM) band using a frequency-hopping technique in which devices change channels at 50 hops per second.

HomeRF-based data networks are expected to have a maximum data throughput of between 0.5 Mbps and 2.0 Mbps, with 1.6 Mbps given as the most likely level. The HomeRF Working Group, established to design an open industry specification, is working to increase the data throughput of HomeRF. However, an FCC approval is required for HomeRF to move to the 3- or 5-GHz band, which is expected shortly.

HomeRF technology has several important benefits: Like wireless networking technologies in general, HomeRF is an elegant connectivity solution that is easy to install and maintain. It enables users to have a large degree of mobility. It is also a very cost-effective solution, particularly when compared with the more expensive 802.11b technology; sample costs hit $500 for Apple's 802.11b-based AirPort Base Station and two networking cards versus $380 for the HomeRF-based AnyPoint USB model and two networking cards. Finally, it holds the potential for voice applications, which HomeRF advocates refer to as their "ace in the hole." For example, one could interface seamlessly using a handset with the network for tasks such as activating commands and voice communications.

Significant problems remain for HomeRF technologies. The technology has not yet gained significant market acceptance, mostly due to a dearth of products from high-profile companies and a lack of applications requiring simultaneous voice and data support. Data throughput is an issue. While 1.6 Mbps should be sufficient for file transfers, Internet access, and multiplayer gaming, transmitting video or other data-intensive applications will be problematic. Another problem is that, as the HomeRF name implies, it is geared toward the home market. With 802.11b positioned by many for both the enterprise and home environments, it will be the choice of many users who want to use the same wireless technology at home as they do on the road or at work.

The introduction of HomeRF products this year should reduce the problem of market acceptance. At the time of publication, Intel had just announced its AnyPoint Wireless Home Networking product. Proxim, S3/Diamond, WebGear, Compaq, IBM, and Motorola are also expected to release HomeRF home networking products this year.

To address issues of data throughput, the HomeRF Working Group (in particular Intel and Proxim) has petitioned the FCC to allow frequency-hopping wireless LANs to operate at data rates of up to 10 Mbps, which will necessitate moving to the 3- or 5-GHz band. Based on initial FCC comments, a ruling favorable to HomeRF that will relax the channel widths for frequency hopping in the 2.45-GHz ISM band was approved in mid-2000. If it receives a favorable ruling, a SWAP 2.0 specification, offering data rates of up to 10 Mbps and backward compatibility, is expected in late 2000. In the opinion of e-Market Dynamics, an unfavorable FCC ruling would more than likely push HomeRF to the sidelines.

Note that, in order to achieve better speed, HomeRF manufacturers will need to cut the maximum power from 1 Gigawatt (W) to either 330 or 200 megawatts (mW). This will also reduce the range of HomeRF devices from a maximum of 1,000 feet to perhaps 300 feet. At the same time, this reduction will increase chances of HomeRF interference with 802.11b equipment. WECA and Bluetooth groups oppose the FCC change, mostly on the grounds that 10-Mbps HomeRF products may interfere with existing products based on 802.11b.

KEY WIRELESS HOME NETWORKING VENDORS

Vendors addressing the wireless home networking space are discussed here:

ALATION SYSTEMS. Alation Systems has developed the HomeCast Open Protocol (HOP), which is used in the S3/Diamond HomeFree wireless networking kit. HOP is specified as an industry-standard protocol with a focus on consumer and productivity applications. HOP is the basis of a 1-Mbps home WLAN solution jointly developed with National Semiconductor Inc. The HOP protocol simply appends a HOP-specific header to an 802.3 packet and hands the data to the radio section, where it is fragmented and reassembled dynamically in response to channel quality. All packets are received by all HOP nodes simultaneously, and all nodes communicate with all other logically connected nodes. Connections are bidirectional, and every node is simultaneously connected to all other nodes in the network (a HOP network resembles a star bus network topology). HOP technology utilizes a 2.4-GHz FHSS radio to deliver data wirelessly, and it allows up to 16 devices to communicate on a peer-to-peer network with a line-of-sight range of 50 to 100 meters, depending on wall and ceiling barriers. This solution is relatively inexpensive, with sub-$100 price points.

APPLE. Apple's AirPort product, which began shipping in December 1999, is based on the 802.11b standard and uses Lucent technology. AirPort is one of the few 802.11b products directly targeted at the residential market. The product is somewhat expensive: The AirPort Base Station lists for $299, and AirPort adapter cards are $99 each, with total costs for a two-node AirPort network reaching $500.

CISCO. Cisco's acquisition of Aironet should raise the visibility of that product line, and it bodes well for the future of wireless networking. PC OEM Dell offers the Cisco Aironet 340 series in a peer-to-peer configuration; PC cards cost $179, and PCI is $139.

INTEL. Intel also has a wireless home networking solution based on HomeRF branded with its AnyPoint brand. It is composed of an internal PC card for $129 and external USB model for $119. It will be bundled in the IBM NetVista PC.

LUCENT. Lucent has become a prominent 802.11b vendor with its Orinoco product line and aggressive promotion, particularly toward vertical markets such as hotels, airports, and convention centers. In February 2000, the vendor recast its WaveLAN wireless LAN products with the Orinoco brand, symbolizing its migration of wireless LANs from a vertical market solution to a horizontal (home, business, and public) solution. Orinoco PC card pricing starts at $179. The Orinoco product lineup also features residential gateways, access points, outdoor routers, and access servers.

PROXIM. Proxim uses OpenAir, a proprietary protocol, which was developed prior to 802.11, using FHSS technology. It offers a less restrictive set of specifications, enabling less expensive radio modems; it is therefore a less expensive technology to implement than 802.11. Proxim has had success in the wireless LAN market with its RangeLAN and Symphony products. Symphony enables users to network up to 10 computers within a 150-foot range through walls and floors. It is a peer-to-peer network (and does not require a hub) and uses the 2.4-GHz radio band. Data transmission is rated at up to 1.6 Mbps, although tests by ZD Labs showed throughput at less than half the rated speed. The Symphony lineup includes PCI cards, Industry Standard Architecture (ISA) cards, and PC cards, as well as a bridge product. The cost per PCI card is $129. Proxim also supports the SWAP specification from the HomeRF Working Group. SWAP-based Symphony products are interoperable with Proxim's OpenAir-installed base, based on Harmony software, which automatically detects whether the OpenAir or SWAP standard is in use and adjusts accordingly to eliminate conflicts.

RADIOLAN. RadioLAN leaped ahead of the rest of the wireless LAN market in late 1996 by developing a proprietary solution, with a MAC protocol based on a wireless implementation of Ethernet (IEEE 802.3). RadioLAN's ISA and PCI CardLINK wireless networking adapters broadcast data packets at 5.8 GHz. RadioLAN is fast, with data throughput around 8 to 9 Mbps, close to the claimed 10-Mbps rating, with an effective range of 120 feet. RadioLAN is relatively expensive at $399 per node.

SHAREWAVE. ShareWave offers a proprietary wireless residential networking solution based on the 802.11b standard in the form of wireless home networking controllers and software. ShareWave has developed the WhiteCap networking protocol—the language spoken by the ShareWave network. WhiteCap's key features include QoS, selectable error correction (SEC), and dynamic TDMA. It enables networks to handle bursty data communications among PCs and PDAs while simultaneously streaming full-motion video to TV and packet-based voice to Internet cordless phones. WhiteCap uses FHSS with a data rate of up to 11 Mbps. ShareWave's technology anticipates and supports multimedia capability; significantly, this approach ties in many more devices, connections, and content, enabling not only multiple-PC but also single-PC and even non-PC homes to benefit from home networking solutions. While ShareWave will not bring a home networking product to market, it does provide the silicon and software, and it is working closely with vendors to bring to market products based on ShareWave technology this year.

BLUETOOTH. Bluetooth, an evolving 2.45-GHz short-range radio technology, has gained wide support as a wire/infrared replacement designed to link and sync devices to one another or to networks. Alternatively, it could be used as a short-range point-to-point network or for connecting networks to other networks. Bluetooth uses FHSS, with 1,600 frequency hops per second to preserve transmission security. Groups representing most of the other home networking technologies are working with the Bluetooth Special Interest Group (SIG) to ensure interoperability. Adding Bluetooth capability to a device currently costs about $30, and Bluetooth-enabled products are expected to be on the market by the end of this year.

Bluetooth technology embedded in PCs allows the formation of a piconet that supports up to eight Bluetooth-enabled devices. As soon as a PC is powered on and connection is authorized, a unique ID code for each of the other devices is recorded on the chip so that future communications can proceed automatically. One device then establishes itself as the master within a piconet, and the other devices in the piconet synchronize to a particular hopping pattern that is distinct from any other hopping patterns in the vicinity. Up to 10 piconets can link together to form a scatternet.

Bluetooth technology should have a place within the context of the home network. It will be able to communicate on a peer-to-peer basis with any other Bluetooth-enabled devices within a 10-meter range. Bluetooth devices are expected to achieve compatibility with most home networks, so data can be exchanged. In its most current iteration, Bluetooth will not be the basis for a full home network but will rather be used by specific devices (such as information appliances and set-top boxes), either bridging into the home network or perhaps even remaining a separate network.

However, an IEEE 802.15 working group is exploring ways to extend Bluetooth's range and speed from 10 meters at 721 Kbps to perhaps 100 meters at 10-20 Mbps. The latter capabilities would be attractive in developing home networking products and would bring Bluetooth into direct competition with other wireless LAN technologies. However, to gain this type of speed and range would probably require a revised specification and a move to the 5-GHz radio band. This is possible as signs indicate that the 2.0 spec (which is expected in about a year) will indeed support a higher range and more bandwidth, as well as a potential move to the 5-GHz radio band.

Advantages of Bluetooth technology include attractive price points, flexibility, and mobility. Bluetooth also has fairly wide support across the industry; over 1,000 companies support the standard. However, significant disadvantages also need to be overcome. Interoperability with other standards, particularly 802.11b, is an issue, and both groups are working to alleviate problems. Bluetooth also has a very limited range of 10 meters; e-Market Dynamics feels that Bluetooth's functionality as a short-range communications tool makes it unnecessary to lengthen that range until (and unless) its capabilities are enhanced. Finally, even though there has been much discussion about Bluetooth, the market has yet to see the introduction of any Bluetooth-enabled products.

WORKING WITH WIRELESS NETWORKING

Benefits of Wireless LANs

The widespread strategic reliance on networking among competitive businesses and the meteoric growth of the Internet and online services are strong testimonies to the benefits of shared data and shared resources. With wireless LANs, users can access shared information without looking for a place to plug in, and network managers can set up or augment networks without installing or moving wires. Wireless LANs offer the following productivity, service, convenience, and cost advantages over traditional wired networks:

- **Mobility.** Wireless LAN systems can provide LAN users with access to real-time information anywhere in their organization. This mobility supports productivity and service opportunities not possible with wired networks.

- **Installation speed and simplicity.** Installing a wireless LAN system can be fast and easy, and can eliminate the need to pull cable through walls and ceilings.

- **Installation flexibility.** Wireless technology allows the network to go where wire cannot go.

- **Reduced cost of ownership.** While the initial investment required for wireless LAN hardware can be higher than the cost of wired LAN hardware, overall installation expenses and life-cycle costs can be significantly lower. Long-term cost benefits are greatest in dynamic environments requiring frequent moves, additions, and changes.

- **Scalability.** Wireless LAN systems can be configured in a variety of topologies to meet the needs of specific applications and installations. Configurations are easily changed; they range from peer-to-peer networks suitable for a small number of users to full infrastructure networks of thousands of users that allow roaming over a broad area.

WIRELESS TECHNOLOGIES OVERVIEW

The essence of a wireless LAN is the cell. The cell is the area where all wireless communication takes place. In general, a cell covers an area that is more or less circular. Within each cell are radio traffic management units also known as access points (repeaters). An access point in turn interconnects cells of a wireless LAN and also connects to a wired Ethernet LAN through some sort of cable connection.

The number of wireless stations per cell is dependent on the amount and type of data traffic. Each cell can carry anywhere from 50 to 200 stations depending on how busy the cell is. To allow continuous communication between cells, individual cells overlap. Cells can also be used in a stand-alone environment to accommodate traffic needs for a small- to medium-sized LAN between workstations and/or work groups. A stand-alone cell would require no cabling.

Another option is *wired bridging*. In a wired bridging configuration, each access point is wired to the backbone of a wired Ethernet LAN. Once connected to a wired LAN, network management functions of the wired and the wireless LANs can be controlled by the access points and the routers associated with them. *Wireless bridging* is also an option; this configuration enables cells to be connected to remote wireless LANs. In this situation, networking can stretch for miles if the wireless access points, acting as signal repeaters, are linked successively.

Finally, when several access points are connected to external directional antennas instead of using their built-in omnidirectional antennas, access points can provide multicells. This is useful for areas of heavy network traffic because with this configuration the subscribers are able to automatically choose the best access point to communicate with. Roaming can also be provided for portable stations. Roaming is seamless, and it allows a work session to be maintained when moving from one cell to another cell (there is a momentary break—imperceptible to users—in data flow).

WIRELESS LAN TECHNOLOGY OPTIONS

Manufacturers of wireless LANs have a range of technologies to choose from when designing a wireless LAN solution. Each technology comes with its own set of advantages and limitations.

SPREAD SPECTRUM

Most wireless LAN systems use spread-spectrum technology, a wideband radio frequency technique developed by the military for use in reliable, secure, mission-critical communications systems. Spread spectrum is designed to trade off bandwidth efficiency for reliability, integrity, and security. In other words, more bandwidth is consumed in spread spectrum than is used in narrowband transmission, but the trade-off produces a signal that is, in effect, louder and thus easier to detect, provided that the receiver knows the parameters of the spread-spectrum signal being broadcast. If a receiver is not tuned to the right frequency, a spread-spectrum signal looks like background noise. There are two types of spread-spectrum radio: frequency hopping, also referred to as FHSS, and direct sequence (DSSS).

Narrowband Technology

A narrowband radio system transmits and receives user information on a specific radio frequency. Narrowband radio keeps the radio signal frequency as narrow as possible just to pass the information. Undesirable cross-talk between communications channels is avoided by carefully coordinating different users on different channel frequencies.

A private telephone line is much like a radio frequency. When each home in a neighborhood has its own private telephone line, people in one home cannot listen to calls made to other homes. In a radio system, privacy and noninterference are accomplished by the use of separate radio frequencies. The radio receiver filters out all radio signals except the ones on its designated frequency.

Frequency-Hopping Spread-Spectrum Technology

FHSS uses a narrowband carrier that changes frequency in a pattern known to both transmitter and receiver. Properly synchronized, the net effect is to maintain a single logical channel. To an unintended receiver, FHSS appears to be short-duration impulse noise. Figure 1.3 provides an illustration of how FHSS works.

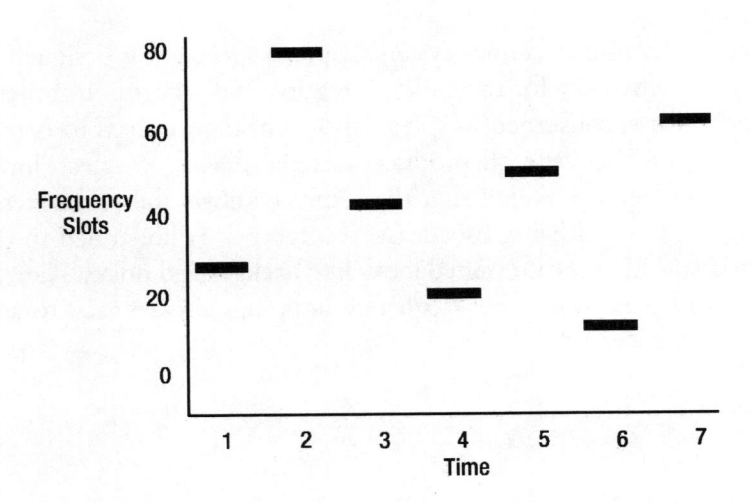

FIGURE 1.2 Frequency-Hopping Spread-Spectrum Technology

DIRECT-SEQUENCE SPREAD-SPECTRUM TECHNOLOGY

DSSS generates a redundant bit pattern for each bit to be transmitted. This bit pattern is called a chip (or chipping code). The longer the chip, the greater the probability that the original data can be recovered (and, of course, the more bandwidth required). Even if one or more bits in the chip are damaged during transmission, statistical techniques embedded in the radio can recover the original data without the need for retransmission. To an unintended receiver, DSSS appears as low-power wideband noise and is rejected (ignored) by most narrowband receivers. Figure 1.3 provides an overview of how this technology works.

INFRARED TECHNOLOGY

Infrared (IR) systems use very high frequencies, just below visible light in the electromagnetic spectrum, to carry data. Like light, IR cannot penetrate opaque objects; it is either directed (line-of-sight) or diffuse technology. Inexpensive directed systems provide very limited range (3 feet) and are used typically for PANs but occasionally are used in specific WLAN applications. High-performance directed IR is impractical for mobile users and is therefore

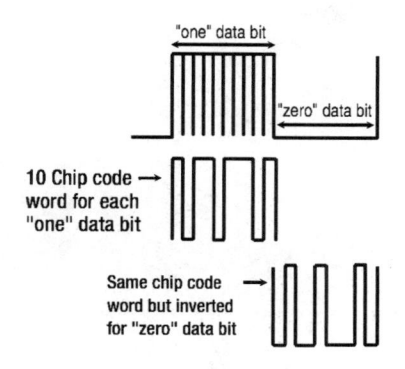

FIGURE 1.3 Direct-Sequence Spread-Spectrum Technology

used to implement only fixed subnetworks. Diffuse (or reflective) IR WLAN systems do not require line of sight, but cells are limited to individual rooms.

MICROCELLS AND ROAMING

Wireless communication is limited by how far signals carry for a given power output. WLANs use cells, called *microcells*, similar to the cellular telephone system to extend the range of wireless connectivity. At any point in time, a mobile PC equipped with a WLAN adapter is associated with a single access point and its microcell, or area of coverage. Individual microcells overlap to allow continuous communication within the wired network; Figure 1.4 provides a graphical example of how this technology works. Microcells handle low-power signals and hand off users roaming through a given geographic area.

WIRELESS LAN

Wireless LANs can be simple or complex. In the most basic configuration, two PCs equipped with wireless adapter cards can set up an independent network whenever they are within range of one another. This is called a *peer-to-peer network*. On-demand networks such as this require no administration or preconfiguration, and each client would have access to the resources of

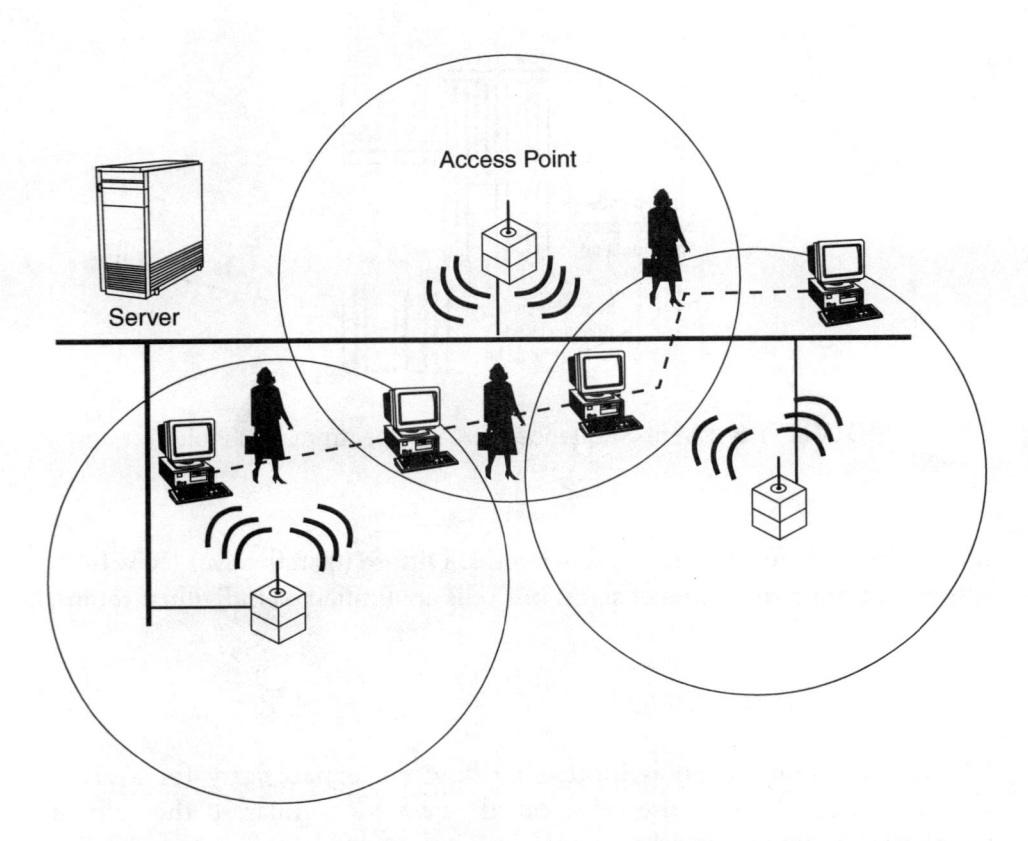

FIGURE 1.4 Handing Off the WLAN Connection between Access Points

FIGURE 1.5 A Wireless Peer-to-Peer Network

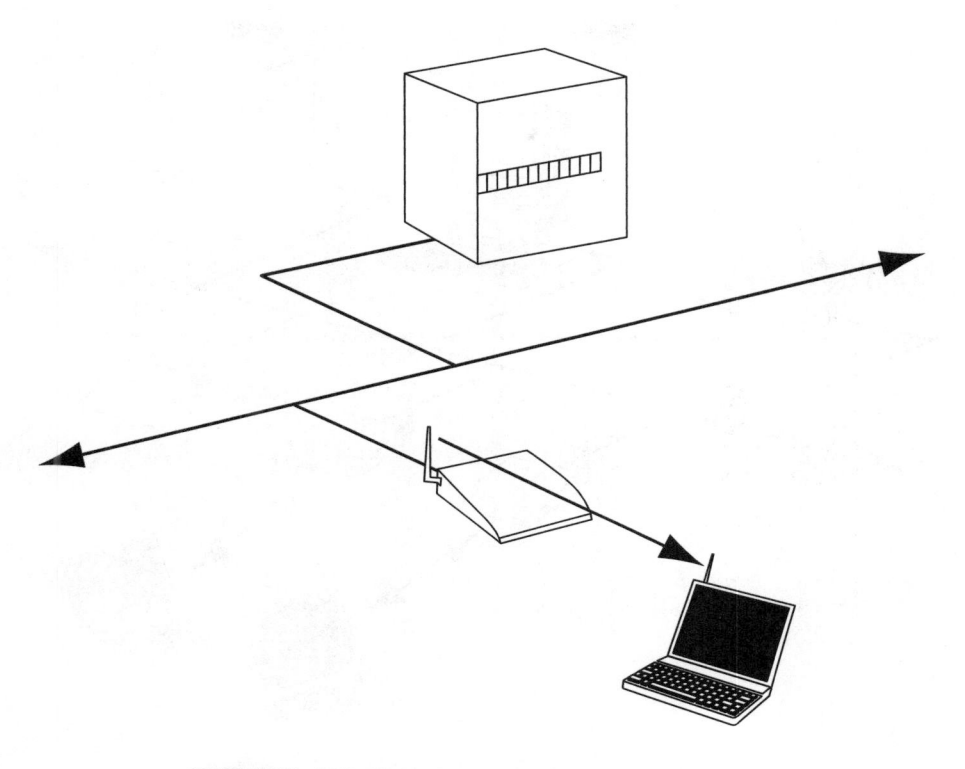

FIGURE 1.6 Wireless Client and Access Point

only the other client and not to a central server. Figure 1.5 provides an illustration of how a wireless peer-to-peer network works.

Installing an access point can extend the range of an ad hoc network, effectively doubling the range at which the devices can communicate. Because the access point is connected to the wired network, each client would have access to server resources as well as to other clients. Each access point can accommodate many clients; the specific number depends on the number and nature of the transmissions involved. Many real-world applications exist where a single access point services 15–50 client devices. Figure 1.6 illustrates a wireless client and an access point device.

Access points have a finite range—on the order of 500 feet indoors and 1,000 feet outdoors. In a very large facility, such as a warehouse or on a college campus, it will probably be necessary to install more than one access point.

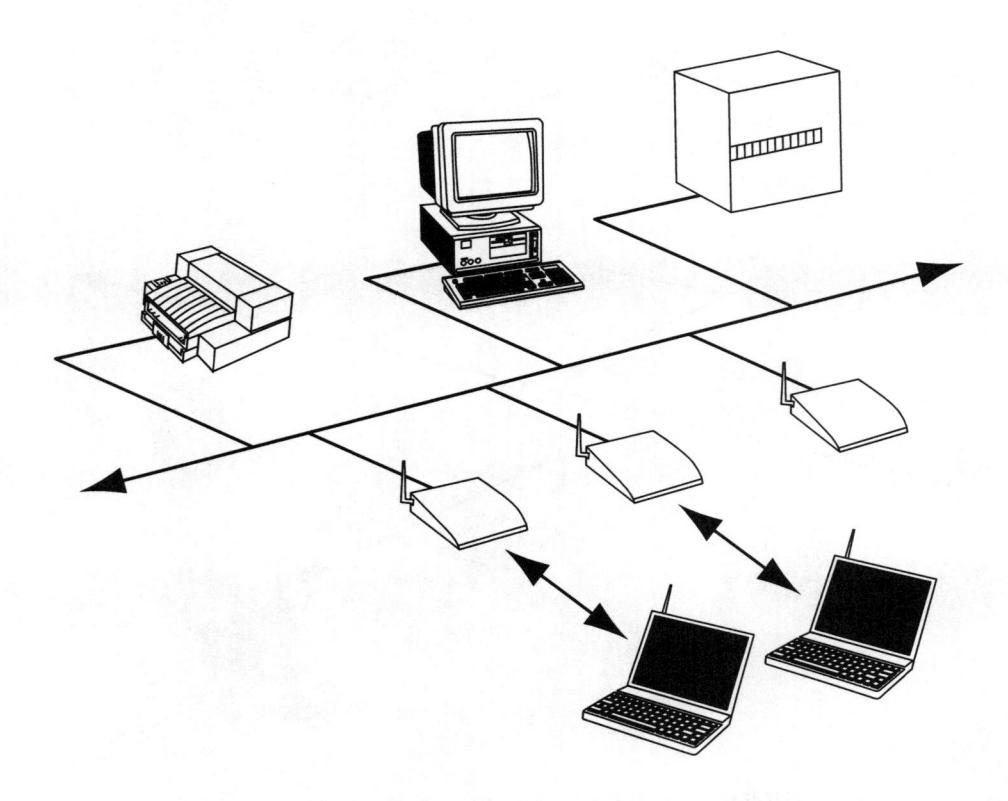

FIGURE 1.7 Multiple Access Points and Roaming

Access point positioning is accomplished by means of a site survey. The goal is to blanket the coverage area with overlapping coverage cells so that clients might range throughout the area without ever losing network contact. The ability of clients to move seamlessly among a cluster of access points is called *roaming*. Access points hand the client off from one to another in a way that is invisible to the client, ensuring unbroken connectivity.

To solve particular problems of topology, the network designer might choose to use extension points to augment the network of access points. Extension points look and function like access points, but they are not tethered to the wired network as are access points. Extension points function just as their name implies: They extend the range of the network by relaying signals from a client to an access point or another extension point. Extension points may be strung together in order to pass along messaging from an access point to

CHAPTER 1: FUNDAMENTALS OF LANS

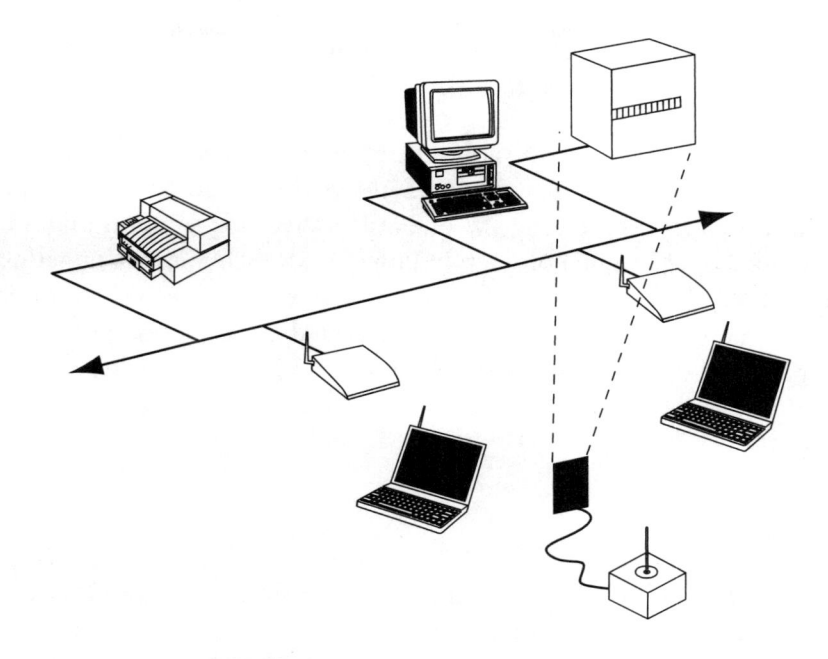

FIGURE 1.8 Use of an Extension Point

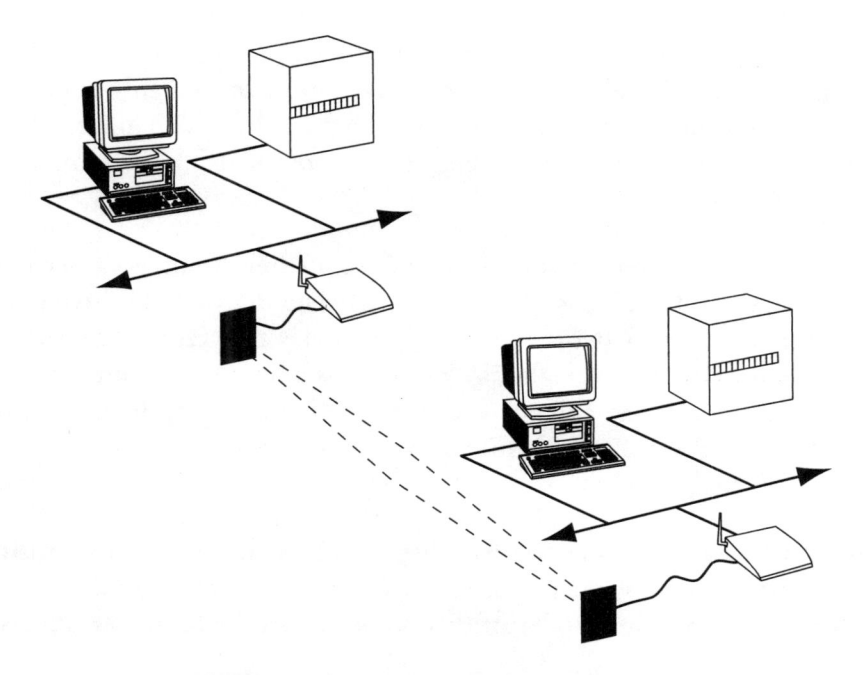

FIGURE 1.9 Use of Directional Antennas

31

far-flung clients, just as humans in a bucket brigade pass pails of water hand-to-hand from a water source to a fire.

One last item of wireless LAN equipment to consider is the directional antenna. Let's suppose you had a wireless LAN in your building (A) and wanted to extend it to a leased building (B) one mile away. One solution might be to install a directional antenna on each building, with each antenna targeting the other. The antenna on A is connected to your wired network via an access point. The antenna on B is similarly connected to an access point in that building, which enables wireless LAN connectivity in that facility.

SUMMARY

Home networks are quickly becoming commonplace in today's households. Many families started with their first PC more than 10 years ago and have grown accustomed to Internet access. The comfort with technology as a part of daily life is evidenced by more than 60 percent of all homes in the U.S. having PCs and nearly 25 percent of those homes with networks up and running today. With wireless technology the percentage of homes with networks is expected to jump significantly, even more so in highly populated areas once the security aspects of this technology are dealt with and resolved. These market figures are from recent studies completed by International Data Corporation.

The fact that the Internet is starting to eclipse television as a form of entertainment for families is a further social dynamic that is driving the development of home networks. With this relatively new reliance on the PC and its shared resources throughout a network as a home entertainment and knowledge center, the acceptance and growth of ancillary technologies, including routers, HomePNA technologies, and wireless, will pervade global homes in the coming years.

Of all home-networking technologies, Ethernet is by far the most pervasive, with wireless showing great promise. Throughout this chapter, and in selected chapters of the remainder of this book, the strengths and weaknesses of Phone Line, Power Line, and the wireless standards HomeRF and 802.11b are explored. There's also the Bluetooth technology, meant for short distances and the synchronization of devices at a local level. You would think that the development of kits—all-inclusive packages that have the correct routers and cabling for each respective technology—would be very popular. In fact, the opposite is the case. More and more enthusiasts are building their networks through the selection of each component for its specific merits. It's common to see many people purchasing a 3COM router, for example, based on the reputation of that manufacturer in this arena. In looking to create your own network, consider it comparable to creating a stereo system; just as you would select a specific receiver, CD player, or a DVD player to hook into your TV, that same dynamic holds true for the world of creating a home network.

The world of home networking can appear complex and difficult to navigate. Don't be intimidated by the jargon and technical terms. Keep the allegory in mind of building a stereo system and you'll do great. Just keep in mind what it is you really want from the network in your home. Is it always-up time and fast connections? The traditional approaches of wiring your home, even using HomePNA, might be best for you. If you have laptops and want to roam the house, using them anytime you want, anywhere you want, then wireless is the way to go. Keep in mind that the security aspects of wireless communication are very much like having a low-frequency portable phone: Both can be overheard. The one drawback of wireless is the lack of security in high-density environments. If you are surrounded by acres of your own land, wireless is definitely for you. If you are surrounded by thousands of other people, either in a subdivision or in a building, then think twice about wireless as anyone else also running wireless has the potential to break into your network.

Lastly, the process of creating a network can be a challenging one, so it's best to go into the tasks here with a sense of adventure and learning. The fact that networks can make your life easier and can streamline communication in the busiest of households makes them worth looking at for that contribution alone.

CHAPTER 2

Choosing a LAN Cabling Solution for Your Home or Office

Case Study: Wireless Networking Quickly Adopted by Ice Cream Maker

At Wells' Dairy, manufacturer of Blue Bunny ice cream, the flavor of the month is wireless. Since beginning the move to wireless in the summer of 2000, the company's information services staff has been inundated with requests from various departments to be added to the network.

Founded in 1913, Wells' Dairy is the world's largest family-owned and managed dairy and the largest copacker of ice cream in the United States. Because Wells' Le Mars, Iowa, plant produces the world's highest volume of ice cream in any one location, Le Mars has become widely known as the "Ice Cream Capital of the World."

Wells' Dairy has installed Cisco Aironet 340 series access points and client adapter cards in deploying numerous internal LANs at its main corporate office, corporate annex, and information systems (IS) center. The access points are wireless, 11-Mbps LAN transceivers that function as bridges between a wireless and a wired network, or as the hub of a stand-alone wireless network.

The client adapter cards enable users of portable equipment to enjoy uninterrupted access to centrally located data while moving freely throughout a specified environment. Desktop computers can also be connected. The Cisco Aironet solution provided PCI cards with 32-bit PCI slots for the dairy's desktop systems and Type II PC cards for the its laptops. An ISA card is also available.

Cisco Aironet 340 series access points and client adapter cards are IEEE 802.11b compliant and feature 128-bit WEP encryption, which provides data security that is equivalent to a wired LAN.

"To be honest, we started with Lucent, but found Cisco to be superior, and we are swapping it all out throughout the entire network," says Jim Kirby, Wells' network engineer. "Several reviews in the network media have given [the] Cisco Aironet [solution] the edge in side-by-side throughput comparisons, and we've tested it ourselves. We've done some hammering on the links with heavy traffic to see what kind of performance we get, and it's definitely superior to Lucent. Some of our users even contend that the throughput with Cisco is faster than hardwired."

Kirby also cited other factors in Cisco's favor: ease of configuration and ease of management. "Management on the Cisco access points is superior to any other network device I've ever worked with. It's the most useful and usable management interface I've ever encountered. There is so much information available on the system."

The ability to broadcast configuration changes networkwide from one point is especially useful. "For example," says Kirby, "we can push out firmware—the software that runs the access point—from any one access point to all the other access points. It works fabulously, and it's one of the reasons that we are now an all-Cisco shop." Wells' Dairy employs Cisco switches, routers, and firewalls.

The Cisco Aironet wireless system has proven itself to be cost effective for Wells' Dairy in several ways. "We are required to run two Category 5 cables from the closet to each station—one each for telephony and network. Each network connection that is cabled to a station requires a switch port. Wireless thus saves us half our normal cabling costs, since there's no data net cabling, and about 30 to 45 percent on installation costs, chiefly in reduced man-hours," explains Kirby.

CHAPTER 2: CHOOSING A LAN CABLING SOLUTION

Wells' Dairy handled the installation themselves using one basic configuration for the access points and one for the adapter cards. Overall, the company found wireless much easier to set up—because technicians are no longer required to install patch cables for every station on the floor, this freed information technology (IT) staff for other jobs.

The move to wireless is progressing smoothly. By the fall of 2000, 25 laptops and 70 desktops had been converted, and the IS staff has another 40 laptop cards and 120 desktop cards ready to install.

Most of the wireless users are in accounting, purchasing, and other corporate positions, including key executives, who regularly carry their wireless-connected laptops to board meetings and such.

Sales personnel, who are frequently on the road, are finding wireless especially helpful. In fact, wireless was first tested in some of Wells' remote, sales-only offices. Those personnel overwhelmingly endorsed it, and now all the sales offices want it. Regional sales offices in Omaha, Kansas City, and Joplin, Missouri, have been equipped with wireless.

Response from the Wells' Dairy users has been overwhelming. "They love it. We can't get it out fast enough. Even people who already have hardwire connections are requesting it," Kirby comments.

USING COPPER WIRING

Phone wiring isn't just for phones anymore. However, ordinary telephone wiring can't handle today's rapidly expanding communications needs. Today's homeowners expect their homes to accommodate the following:

- Multiple phone lines
- Internet service
- Video distribution, and other entertainment services
- Data and security services

- Fax machines

- And the list goes on . . .

Every room in the modern home should have low-cost, high-tech copper wiring (Category 5 or better), which is faster and more reliable than ordinary phone wiring. That's what is needed to carry voice, data, and other services from their entry point into the house to every room *and* from any one room to any other room.

This section will give you the basics on home wiring for the rapidly evolving information age.

UTP copper information wiring—often called *structured wiring*—is used today for LANs in offices, schools, and factories, allowing computers to talk to one another and to receive and send Internet and high-speed computer data outside the facility. Category 5 (or Cat 5) is the current standard, but will soon be supplanted by even higher-speed versions, known as Category 5E (Cat 5 enhanced) and Category 6. (Cat 6 has at least twice the bandwidth, or information-carrying capacity, of Cat 5 at a small cost premium.) Right now, the typical home doesn't require the capacity to move computer signals around as fast as the typical office. However, offices get extensively remodeled—and rewired—every few years; homes do not. The wiring installed in a home must serve indefinitely.

The phone wiring of the past, often referred to as *quad wiring* because it has four copper wires, is now obsolete. Cat 5 or higher-speed wiring has four twisted wire pairs, or eight wires. All are needed to provide the multiple services discussed here.

In fact, an FCC ruling, effective July 2000, now *requires* that homes as well as businesses be wired for the information age.

Copper UTP Wiring: What Is It?

Copper UTP wiring contains eight color-coded conductors (four twisted pairs of copper wires). It offers greatly increased bandwidth in comparison to old-fashioned quad wiring. The cable is small (roughly 3/16 inch in diameter), inexpensive, and easy to pull, although it must be handled with care.

Modern copper UTP wiring offers the following advantages:

- **Diversity.** The Internet and computer communications, as well as ordinary phone signals, can be carried throughout the home on modern, inexpensive, high-speed UTP cables. (To service a large number of TV channels, it is recommended to also run high-quality coaxial cable, such as quad-shielded RG-6.)

- **More phone numbers.** Several phone numbers can be made available throughout the house. Actually, voice service requires very little bandwidth, and the addition of separate numbers is almost trivial.

- **Bandwidth.** Cat 5 has an approved bandwidth of 100 MHz, while Cat 6, when finally approved as a standard, will likely accommodate at least 200 MHz when tested under stringent conditions. Bandwidth correlates with speed, and these bandwidths are *many* orders of magnitude *greater* than the bandwidth required for a "modern" 56-kbps (kilobits per second) modem. Category 6 wiring, with encoding, will be able to carry at least 1 gigabit (billion bits) per second (gbps). If you're counting, that's about *50,000 pages* of text per second!

- **New Services.** The Internet is now available at high speed to many homes, but homeowners won't be able to take full advantage of it if their wiring is inadequate. One high-capacity technology now being offered by local phone companies is digital subscriber line (DSL). And cable modems are being offered by cable TV companies that bring the Internet in on the same coaxial cable that carries the TV signals.

- **Reliability.** Interference on telecommunications lines can result in scrambled faxes, interrupted online sessions, and distorted video and audio signals. High-tech twisted-pair copper wiring—due to its tight, accurate twist of the wire pairs and its balanced mode of transmission—is designed to resist interference from sources in the home, such as microwave ovens and other appliances, vacuum cleaners, fluorescent lights, power tools, and external communications signals.

- **Approved.** Performance of these cables is verified by Underwriters Laboratories (UL), the international product testing agency, and similar groups: UL, TUV, and CSA–all European standards organizations that are very important to in-home use. FCC Class B in the United States equates to the business system ergonomic standards in Europe, as many of the countries there have a heightened sense of the environmental impact of computing and networking equipment.

To provide the home with the most modern, flexible voice and data services, follow these four simple principles when installing telecommunications wiring in a residence:

1. Use Cat 5 (or better) UTP copper wiring.

2. Wire every room.

3. Use a star wiring pattern.

4. Use 8-pin modular jacks.

Here's why:

- **Use Cat 5 (or better) UTP wiring.** With this wiring, you can do the following: move large quantities of data around the house; move a limited number of TV signals from the point of entrance to anywhere and everywhere else in the house (using a readily available adapter); move other signal-level entertainment to as many locations as desired; connect phone, fax, and computer/printer wherever desired.

- **Wire every room.** Because you can only guess which rooms in the future will be used for what when you wire a house, it's usually best to provide outlets virtually everywhere. For instance, the kitchen is often the business center of a household and thus needs multiple jacks.

- **Use a star wiring pattern.** With star wiring, each outlet (jack) has its own individual *home run* of cabling extending back to a central distribution device. There are three major advantages to this:

- *Flexibility*—all changes in distribution of services can quickly and easily be made at the central distribution device, and each outlet can be treated independently (in loop wiring—also known as *daisy-chain* wiring—where a number of outlets are tied together in series, outlets cannot be treated independently).

- *Isolation of problems*—when an interruption takes place (nail through a wall and into a cable, etc.), only one outlet is affected.

- *Quality of signal*—each additional connection point is a potential source of interference and other problems that can cause a loss of signal quality.

Putting an extra outlet or two in some rooms, particularly home offices, is a wise move, as you can't anticipate future room use or furniture arrangements. Do this with home runs to the additional outlets. Most seasoned professionals strongly recommend running extra "wire" to any location where it might be needed later. For example, you might run two 4-pair cables, rather than one to each outlet, to enable expansion and flexibility. To further future-proof your homes for only a small additional cost, consider running Cat 5E or 6 all the way, but particularly for an area that might be used as a home office.

USING 8-PIN MODULAR (RJ-45)

These devices provide connection points for all eight of the wires contained in the four twisted pairs. Figure 2.1 shows a wall outlet with two such jacks.

Figure 2.1 How Connections Work with Twisted Pairs

A contractor needs to understand communications wiring and to know how to install it according to strict guidelines, such as the following:

- How much pull can be applied to a cable (usually 25 pounds)

- How much separation is needed between data and power cables (6 inches at least, crossing, if necessary, at 90 degrees)

- What fixtures to avoid (fluorescent lights in particular)

- How far back the cable sheathing can be stripped (no farther than necessary, typically 1 1/4 inches)

- How much untwisting of the pairs can be done when making connections (1/2 inch is usually the maximum recommended; 3/8 inch is better)

- How tight a bend radius can be tolerated (usually about 1 inch, although some designs are less sensitive than others)

- How long a cable run can be (about 300 feet)

Important: All connecting devices—the central distribution device, plugs on the ends of cables, outlets, etc.—should be rated for the cable used. For instance, if Cat 5 cable is used, all devices must be at least Cat 5 rated; if Cat 5E or 6 cable is used, all devices should be rated Cat 5E or 6.

Finally, the finished installation should be thoroughly tested.

WORKING WITH VIDEO CABLES

Although the industry is working toward an all-UTP solution for wiring residences, at this time it is prudent to also include conventional coaxial cable (or coax, for short) for video distribution, particularly for cable TV. This is because it is difficult to know whether the prediction of a vast array of channels—say, well over 100, some of which will be the more bandwidth-consuming high-definition television (HDTV)—will become a reality in the near future. If coax is installed, quad-shielded RG-6 coax, with an all-copper center conductor, will give superior performance. (Copper-plated steel-center conductors are also available, providing additional stiffness, but are unable to handle the low-frequency currents used to power some devices.) Do not use the lower grade RG-59 coax.

INFRARED

The notion of portable computing takes into account any device (laptop, palm pilot, etc.) that can be carried easily and can be connected to a network. As such, it has become synonymous with the rapidly changing world of technology in today's work environment. Most companies recognize that in order to compete and maintain their competitive edge, they must keep pace, as well as deploy and manage this new technology. New economic trends in the global economy will continue to push companies to find new ways to enhance productivity and maintain flexibility among their employees.

In fact, according to recent studies, portable computing is poised for some very dramatic growth. Many estimates suggest that, when the final numbers are in, the number of portable computers that were sold in 2000 will be double the 9.4 million units sold in 1997. This trend is likely to continue as companies deploy their workforces to a more mobile office concept. One obstacle to achieving the goal of mobile connectivity has always been the limitation of the cable connection. Whether you are connecting to the printer or your network or exchanging data with the desktop, the cable connection is a hindrance to the effective and efficient use of the portable computer.

THE ADVANTAGES OF INFRARED

IR technology was born out of an effort to achieve a wireless connection to a full range of peripheral devices without the hassle of cable.

Some of the benefits of IR technology include:

- A worldwide standard for wireless connectivity

- Easy to implement and simple to use
- Safe in any environment
- No electromagnetic noise
- No government regulatory issues
- Minimum cross-talk

The IrDA Standard

In 1993, leaders from both the communication and computer industry formed the Infra-red Data Association (IrDA) with the sole purpose of creating a standard for IR wireless data transfer.

Today the IrDA association has over 120 members worldwide, including some of the most recognized companies in the world, such as Apple, AT&T, ACTiSYS, Canon, Compaq, Hitachi, Intel, Hewlett Packard, Microsoft, Motorola NTT, Sony, and Toshiba.

Connecting with IR

When two IR devices come within range of each other, Windows95, Windows98, Windows NT, Windows 2000, and Windows XP will automatically detect the device and display its signature on the screen. An audible alert will also sound, indicating that a connection has been made. If for any reason the beam is interrupted, Windows95 will again signal audibly and attempt to reestablish the link for up to 45 seconds. No data loss will occur if the link is reestablished at that time.

Summary of Infrared

In 1995, only 38 percent of the leading portable computer manufacturers shipped IR with their products. By early 1997, that number was expected to reach 100 percent. The reason is simple: Cost, reliability, and flexibility have all contributed to the overwhelming acceptance presently enjoyed by the IR standard. Users now have the option to be free of the hassle and clutter of cables and various connectors, while still having virtually unlimited connectivity to a wide range of desktop and peripheral devices.

A Commitment to Infrared Technology

ACTiSYS's commitment to IR technology dates back to 1989. In September 1993, ACTiSYS was one of early members of the IrDA standards association, and the company remains active to this day. In November 1993, at the Fall Comdex, they introduced the world's first dual-mode IR serial adapter. Today, ACTiSYS boasts one of the most complete lines of IrDA products in the industry, including IR Protocol and test software as well as a full range of IR adapter products.

Appearing in many PowerBook models and even early iMacs, IR networking allowed for wireless, line-of-sight connection between computers, personal digital assistants (PDA), and anything else equipped with an IR (or IrDA) port.

IR is quick and easy to set up, allowing for LocalTalk speeds (cumbersome for modern networks, but acceptable for small file transfer). Portable devices are the primary users of IR, since it eliminates the need to run wires for what is often a temporary connection.

CHAPTER 3

INTRODUCTION TO TCP/IP AND NETWORKING IN WINDOWS 2000

INTRODUCTION

In creating the Windows 2000 operating system, Microsoft began, and has continued, a commitment to listening to its customers, especially the ones in larger network configurations. Microsoft could clearly gain a presence in mid-range to high-end networking in organizations. Resoundingly, these customers said that, in the world of connectivity, Transmission Control Protocol/Internet Protocol (TCP/IP) is a definite requirement for any new operating system, especially those being targeted at the Internet. While this feedback was first obtained when Microsoft began developing the Windows NT operating system over six years ago, the depth of TCP/IP support has undergone significant changes in the last two years. Microsoft learned and capitalized on the fact that TCP/IP is the pavement of the information traffic flow between companies and even continents. Indeed, TCP/IP is the backbone of the Internet itself. Realizing that experience is the best way to learn about TCP/IP, Microsoft integrated 60,000 nodes within its own company to learn more about this protocol. Managing or administering such a large network proved daunting, which led Microsoft to appreciate the need for powerful tools that would make system administration easier than before.

From the knowledge gained through experience, Microsoft built and refined the TCP/IP protocol implementation in NT 4.0.

In getting this large TCP/IP project moving, Microsoft learned about the needs of TCP/IP users in its own company by actively beta-testing both Windows 2000 Professional and all levels of Windows 2000 Server, including the connectivity and specifically commanded by TCP/IP. From this extensive testing and integration effort, features were added to the NT TCP/IP protocol suite including Dynamic Host Configuration Protocol (DHCP) and the Windows Internet Naming Service (WINS).

Now TCP/IP is considered the great equalizer, making it possible for network-based devices of all types to share all types of data, regardless of the operating system of the specific node. With TCP/IP, in conjunction with the Java Virtual machine (JVM) software included in many network browsers, there is a new ubiquity to information-sharing not seen in the previous generation of operating-system-bound devices. Let's get started in this chapter with an overview of the TCP/IP protocol and move through the DHCP and Domain Naming Services (DNS) areas of NT's implementation including Windows NT and Windows2000/WindowsXP.

A BRIEF HISTORY OF TCP/IP

Originally developed to meet the needs of universities, researchers, and even defense contractors to share data across broad geographic distances quickly, TCP/IP has progressed into an ever-increasing range of protocols. ARPANET was originally created as an experiment in large-scale packet switching. The foundation of today's Internet was created due to the goals of the ARPANET project. First created as a research endeavor, the TCP/IP protocol is not slanted toward a given vendor or manufacturer. This actually made the ARPANET more ubiquitous and more nimble at responding to the needs of its users—the growth of the network could draw on the resources of multiple contributing companies and universities, and it did not have to be constrained by the technology of a single provider. ARPANET and now the Internet are successful because they take the best of multiple sources of technology rather than just a single source of innovation.

HOW TCP/IP FITS INTO MICROSOFT'S NETWORKING STRATEGY

The role of TCP/IP in the product strategy of Windows 2000 is to create a foundation within each of the Windows operating systems from the Windows NT 3.51 release and beyond. This means that, as of now, several generations of Windows have been used to translate users' needs into a TCP/IP command set robust enough for the needs of corporate users interested in ensuring connectivity between wide numbers of users.

Microsoft continues to use added functionality and differentiation within the TCP/IP command set to add value to the Windows 2000 product family. Sooner or later you will encounter the need to configure TCP/IP in Windows 2000, and the steps in this chapter will provide you with the necessary tools to do this. The role of TCP/IP in Microsoft's product strategy is to accomplish a variety of tasks. First, TCP/IP provides a solid foundation on which to build a connectivity strategy and drive the development of mixed or heterogeneous networks that include many different operating systems having TCP/IP in common as their method for communicating. Second, TCP/IP plays the critical role of streamlining the integration of Windows 2000 Professional workstations and servers with the Internet. If you are a system administrator, you will be faced with the task of ensuring connectivity between servers and the Internet either via routers and T1 lines from your own organization or via Internet Service Provider (ISP) connections. Third, the customization of the TCP/IP command set by Microsoft for the Windows 2000 operating system specifically creates a differentiator that positions Microsoft effectively versus UNIX and NetWare as a viable alternative.

TCP/IP BASIC CONCEPTS

It was with foresight that the first companies began building TCP/IP into their organizations. UNIX provides TCP/IP as a standard feature, working to ensure that file transfers and connectivity is strengthened. UNIX was the first major operating system to include TCP/IP; due to this operating system's ability to handle multiple sessions, as well as its appeal to the technical applications areas, UNIX and TCP/IP quickly became market standards. Why does TCP/IP continue to gain in acceptance? It's because the founding concepts continue to meet needs and add value as the ongoing need for connectivity requires that the protocol to be the first priority and the operating system secondary. The following sections detail the reasons why TCP/IP continues to gain in popularity within the NT-user community.

Hardware Independence

TCP/IP is the great equalizer when it comes to processors and network connectivity hardware. Instead of being limited to only a certain microprocessor, TCP/IP is truly hardware independent. The growth of Network Computers (NCs) relative to lower-priced traditional PCs points to the fact that the microprocessor, in specialized network applications, becomes secondary. The network protocols will play the role of the operating systems in many cases of these specialized devices. Nowhere will this become more prevalent than on the Internet.

Standardized Addressing

Because any host computer on any TCP/IP network can be individually and specifically addressed, applications and programs using a specific IP address will be able to find the targeted system, and you can know without a doubt that the data sent is intended for that specific system. This standardized addressing technique in TCP/IP also ensures that the assignment of IP addresses will be consistent in meaning and structure, regardless of whether they are assigned through static means with each individual system getting a discrete, separate identity, or assigned as entire pools of PCs being able to draw from a range of addresses. Later this chapter will discuss Dynamic Host Configuration Protocol (DHCP). The essence of this protocol is the creation of an entire pool of IP addresses that are checked in and out in much the same way books and tapes are checked out of a library. In effect, a reservation can be made to get an IP address, and, within the reservation, the time the IP address will be needed will also be communicated. Many of the ISPs that make their revenues based on offering access to the Internet use DHCP for assigning IP addresses to those customers who want access to the Internet. The standardized approach to structuring IP addresses allows for the flexibility of creating protocols such as DHCP that make it possible to dynamically assign addresses from a range or pool instead of having to construct addresses on a one-by-one basis.

OPEN STANDARDS

TCP/IP continues to gain enhancements due to the fact that it is based on an open standard that is available to anyone. The standards organization that manages the TCP/IP protocol is relatively approachable and actually publishes proposed additions to the standard in the form of Requests for Comments (RFCs). RFCs are actually descriptions of planned enhancements to the standard, and anyone is welcome to comment on their content. Some of these documents are quite technical, but many are understandable for someone with a solid understanding of the Open Systems Interconnect (OSI) Model, the essentials of TCP/IP, and with the patience to read through the RFCs. So TCP/IP's open standard is truly that—unlike other highly technical standards, you don't need to be invited to comment; you can...today. You can find these RFCs at the following anonymous File Transfer Protocol (FTP)sites on the Internet:

    DS.INTERNIC.NET

    NIS.NSF.NET

    NISC.JVNC.NET

    FTP.ISI.EDU

    WUARCHIVE.WUSTL.EDU

    SRC.DOC.IC.AC.UK

    FTP.NCREN.NET

FTP.SESQUI.NET

NIS.GARR.IT

APPLICATION PROTOCOLS

Another force driving the increased adoption of TCP/IP is the inclusion of applications that are based on this protocol that make sharing, transferring, and managing files on dissimilar systems possible. For example, you could use the ftp command to transfer files from a mainframe running TCP/IP to a Windows NT workstation, where a spreadsheet would complete the analysis. After finishing the spreadsheet, you ftp the file to a UNIX system where a time series could be completed using the telnet command. All this could be accomplished from the Command Prompt window of the Windows NT workstation you're using. You can easily trade files with mainframes, Macintoshes, and UNIX workstations of any type of operating system—just as long as the TCP/IP application suite is present, the type of operating system is immaterial. TCP/IP, as you can see, levels the playing field between microprocessors as well.

INTRODUCING THE OSI MODEL

In learning about network protocols and how they compare, it's useful to have a frame of reference or a structure to apply to the myriad of terms and concepts. The OSI Model provides this framework for illustrating the components of the TCP/IP networking protocol. While the OSI Model is widely used for showing the differences between network commands and protocols, it's important to realize that there are just as many variations in how protocols are structured as there are protocols, so the OSI Model is a good foundation on which to learn. Just like foundations of houses, the elevation can be quite different for each one. It's the same with network protocols—all are based on the foundation of the OSI Model, yet their elevations or appearance look different based on the needs of the given set of customers and their needs. In this section, first you'll get an overview of the OSI Model, followed by an explanation of how the TCP/IP protocol fits into this structure. You'll also see how the OSI Model is organized to provide for data packets or datagrams to traverse the levels of the model to ensure communication and data reliability.

The OSI Model is organized into a hierarchy where data travels from the lowest point to the top, with each level either stripping away data elements needed for handling the data transaction or, in the case of an outgoing message, adding data to ensure the targeted system receives the complete message intended. The layers of the OSI Model, beginning with the bottom layer and moving upward, are the following:

- Physical layer
- Data link layer
- Network layer
- Transport layer
- Session layer
- Presentation layer
- Application layer

Figure 3.1 shows the layers of the OSI Model, with a brief description of each layer.

The OSI Model provides a useful framework for comparing network protocols. The first two layers—the physical and link layers—define the way data is physically transferred on the network. The network, transport, and session layers define the way data is processed before being passed to the operating system (or processed before being passed to the lower layers of the OSI Model). The presentation layer represents the operating system, and the application layer represents the workstation's network software. Above the application layer are the user interface and general applications that use network resources.

By breaking the network model into different layers, it's also much easier to see how independent and yet linked the physical attributes of network connectiv-

Application Services	**Layer 7 (Application)** - Communications-related services oriented towards specific applications. Examples include file transfer and email.	
	Layer 6 (Presentation) - Negotiates formats, transforms information into agreed-upon format, generates session requests for service	
	Layer 5 (Session) - Manages connections between cooperating applications by establishing and releasing sessions, synchronizing information transfer over these sessions, mapping session-to-transport and session-to-application sessions.	
Networking	**Layer 4 (Transport)** - Manages connections between two end nodes by establishing and releasing end-to-end connections; controlling the size, sequence, and flow of transport packets; mapping transport and network addresses.	
	Layer 3 (Network) - Routes information among source, intermediate, and destination nodes; establishes and maintains connections, if using connection-oriented exchanges or protocols.	
	Layer 2 (Data Link) - Transfers data frames over the physical layer; responsible for reliability.	
Transmission	**Layer 1 (Physical)** - Mechanical, electrical, functional, and procedural aspects of data circuits among network nodes.	

FIGURE 3.1 Overview of the OSI Model

ity are relative to their logical counterparts. For example, the network protocols (TCP/IP, NetBEUI, and so on) are not dependent on the NIC or cabling, but instead work on any hardware because the intervening layers translate and process the network traffic into the format these protocols can understand. Layering the components of the OSI Model also helps make the network functionality transparent to the user interface and applications. Because the layers hide the protocols and physical hardware from the interface and applications, you can change protocols and hardware without making any changes to the interface or to your programs, to enable them to access network resources. Let's take a look at each layer of the OSI Model in more detail, beginning with the bottom layer.

Physical Layer

The lowest layer, the physical layer, defines the connection of one system to another via the network. At this layer, the network transmission consists of individual bits; the upper layers of the network process and package the bits into packet form. The physical layer does not define the media used, but it does dictate the methods by which data is transmitted over the network.

Data Link Layer

The primary purpose of this layer is to process *frames*, which are units of data by which the layer communicates upward in the OSI Model. A frame comprises a string of bits grouped into various fields. In addition to the actual data bits, the frame includes fields to identify the source and destination of the frame, and an error control field that enables the receiving node to determine if a transmission has occurred. The technique used for checking to see if the frame has been received successfully relies on cyclic redundancy checking (CRC), where the data bits are compared using a parity-checking algorithm. This technique of ensuring data integrity was originally developed for hard disk drive controllers.

The data link layer converts outgoing frames into individual bits for transmission, and reassembles incoming bits back into frames. In addition, this layer establishes node-to-node connections and manages the transmission of data between nodes, as well as checking for transmission errors.

Network Layer

The network layer is primarily responsible for routing data packets efficiently between one system and another system on the network. The two network protocols included with Windows NT—TCP/IP and IPX/SPX—are most commonly used in networks where routing is required. The IP protocol, which is part of the TCP/IP suite, is responsible for routing at the network layer. IPX, which is typically used in Novell NetWare environments, serves the same purpose. Microsoft and other vendors involved in network operating systems recognize the necessity of providing routable protocols with their implementations of TCP/IP.

In addition to TCP/IP and IPX/SPX, Windows NT also includes the NetBEUI protocol. In the first versions of NT (3.1 and 3.51), NetBEUI was included as the default network protocol. TCP/IP became the default protocol with version 4.0. The reason for this change was that NetBEUI was originally designed to provide efficient transport in small networks, but it can't be routed. Because large, complex networks often require routing, Microsoft had to change to a protocol that could provide routing as a default for Windows NT—TCP/IP. If you have a small, self-contained network, NetBEUI should be your protocol of choice—or at least one to consider in a peer-to-peer-oriented network. However, if you are going to route data from one location to another, you'll find TCP/IP the best choice for linking dissimilar systems. Or IPX/SPX may be your best choice if your entire network (or the majority of it) is based on Novell NetWare.

Transport Layer

Although the network layer is responsible for routing, it doesn't perform error checking; it simply passes the packets back and forth. There's no guarantee that the packet will arrive with all its components, or, if there are contiguous packets, that they will all arrive in the proper order. Some level of error checking is required, and the transport layer provides this reliability. The transport layer checks for errors and causes the packets to be retransmitted when there is an error in sending or receiving. In essence, the transport layer acts as the first point of quality assurance in the definition of a transaction over the network.

The two protocols that are usually used with Windows2000 because of their routing characteristics are TCP/IP and IPX/SPX. These two protocols have additional features at the transport layer that ensure reliable transmission: The TCP portion of the TCP/IP protocol checks for errors and places packets in the correct order; SPX performs the same function in the IPX/SPX protocol.

Session Layer

This layer is also responsible for adding reliable and secure communication between nodes of the network. When two network nodes or systems communicate, it's called a *session*. The session layer is responsible for establishing the communication link and the rules by which the session will be accomplished, including the transport protocol to be used. This layer also involves security authentication—two server nodes authenticate the client node's request through the security subsystem, determining what the client can and can't access.

Presentation Layer

This layer functions as an intermediary between the session layer and the application layer. The presentation layer translates the data from one format to another if that is what's required for the session and application layers to communicate. The presentation layer also performs such tasks as data compression and encryption.

Application Layer

This layer provides the interface between the network and user interface and applications. In general, the application layer provides a layer of abstraction between the user and the network, making the network essentially transparent. Because of the application layer, a program you're using to open a file can do so as easily across the network as it does locally.

UNDERSTANDING HOW TCP/IP ENABLES COMMUNICATION

The OSI Model provides a useful reference point for comparing the components of the TCP/IP command set. You may be asking yourself, "How will this help me serve my customers regarding TCP/IP?" The answer is that having these foundational elements will enable you to assist and guide them through the most challenging aspect of setting up a network—getting systems to work in conjunction with each other. Three processes are critical if datagrams are to be routed from the source computer to the to the destination host via the intended or destination network:

- **IP Addressing.** With an IP address, each network or workstation has a unique identity and can be specifically communicated with using its specific address.

- **Routing.** This is the process by which messages pass through gateways to reach the destination workstation or network.

- **Multiplexing.** Multiplexing is the process by which data is forwarded to a compatible protocol and port, out of the many running on a network.

IP ADDRESSING

This attribute gives a network, workstation, or even an intranet its own identity within a large extranet or even differentiated from the Internet. It's been called the biggest single differentiator in the entire TCP/IP protocol suite. Many organizations prefer that each of their workstations have its own IP address. Typically called *static IP addressing*, this makes a lot of sense, especially when the workstations will reside permanently at their locations. In contrast, DHCP is a network protocol specifically developed for assigning IP addresses in much the same way a library dispenses books—on demand. Both static IP addressing and DHCP use the same values for the IP addresses; there are just different ways of assigning these values to workstations.

An IP address consists of 32 bits, notated as four decimal values representing one octet each, separated by periods. An octet can have any value from 0 to 255, although certain address values are not used, having been reserved for other uses. Using this approach, it is possible to apply an IP address to an individual workstation or to an entire network.

If you are running TCP/IP on a network that is not linked to the Internet, you can give your networks and hosts any valid IP address values you wish. However, computers that you will be connecting to the Internet must be on a network with an address that has been registered with the Internet Network Information Center, or InterNIC. InterNIC is a clearinghouse for all Internet IP addresses; it is responsible for ensuring that host addresses are all unique.

However, InterNIC does not register individual workstations, only networks. As a system administrator, use the series of IP addresses specific to your organization to assign a unique address for each workstation. The network address is part of the 32-bit IP address. The address class and the subnet mask determine which part of the address represents the network and which part represents the host.

IP Address Classes

InterNIC uses three classes of IP addresses, designating the specific class by the first 3 bits of the address. These classes vary depending on needs of the organizations receiving them. The classes of IP addresses are described as follows:

- **Class A.** The first bit of a class A address is always 0, meaning that the first octet of the address can have a value between 1 and 126. Only the first octet is used to represent the network, leaving the final three octets to identify 16,777,214 possible hosts.

- **Class B.** The first 2 bits of a class B address are always 1 and 0, meaning that the first octet can have a value between 128 and 191. The first two octets are used to identify 16,384 possible networks, leaving the final two octets to identify 65,534 possible hosts.

- **Class C.** The first 3 bits of a class C address are always 1, 1, and 0, meaning that the first octet can have a value between 192 and 223. The first three octets are used to identify 2,097,151 possible networks, leaving the final octet to identify 254 possible hosts.

What Is Subnet Masking?

If you have a network address registered with InterNIC, then the host to whom you assigned that address can be used on the Internet. What if you want to set up numerous internal networks? Do they all need to be on the same InterNIC address? The answer is no; you can use subnet masks to differentiate other networks from the one visible to the Internet. Subnet masks are used for dividing your assigned network into subnetworks and assigning IP addresses based on the physical layout of your own internal network.

A *subnet mask* is a filter designed to register a value in the TCP/IP stack that tells the system which bits of the IP address represent the network and which represent the host. The subnet mask value is expressed in the form of a normal decimal IP address, but it is easier to think of it in its binary equivalent. A subnet mask is 32 bits in which 1s indicate the part of the address representing the network and 0s representing the host. For example, if you have a small, one-segment network and you obtain a class-C address, then you can use it without modification. You would enter a subnet mask of 255.255.255.0 in all of your workstations and assign individual host values using the values of 1 to 254 in the last octet of the IP address. The 255s represent binary octet values of 11111111, meaning that all of the first three octets are used to identify the network.

Here's a more complicated example. Say you are assigned a class B address and you have an internetwork of your own that is composed of many individual segments. It would be impractical to use the class B address as it is (with a subset of 255.255.0.0) and the number of hosts in a single sequence from 1 to 65,534, using the last two octets. Instead you would establish subnets that represent the physical segment of your network. You would do this by applying the subnet mask of 255.255.255.0 on all of your hosts. The first two octets of their IP addresses would be values assigned by InterNIC. The third octet, however, would now represent part of the network address instead of the host address. You can assign values to this third octet representing each of the segments on your network, and then use only the fourth octet to address as many as 254 hosts on each subnet. Subnet masking can get even more complicated than this—a mask can be applied only to certain bits of an octet. Careful conversion between binary and decimal values is then necessary to ensure accurate configurations.

Subnet masking is something that must be organized on a internetwork-wide basis. Be sure to check with your system administrator for the values of the subnet masks you should include.

IP ROUTING

The other aspect of getting messages delivered over a network is having the IP routing in place to ensure that datagrams are delivered to the correct workstation or network. When the source and destination workstations are not on the same network, then one or more gateways are used to route packets to the correct network.

Every TCP/IP system maintains an internal routing table that helps it to make routing decisions. For the average host computer located on a network segment with only a single gateway, the routing decisions are simple: Either a packet is sent to a destination host on the same network, or it is sent to the gateway. As stated earlier, IP is aware of only the computers on the networks to which it is directly attached. It is up to each individual gateway to route the packets onto the next leg of their journey to the destination.

If a host computer is located on a network segment with more than one gateway, packets addressed to other networks are initially all sent to the default gateway that is part of the host configuration. However, the default gateway may have routing information that is unavailable to the host, so it will send an Internet Control and Messaging Protocol (ICMP) Redirect packet instructing the host to use another gateway when sending to a particular IP address. This information is then stored in the host's routing table. When the host attempts to send future packets to that same IP address, it first consults its routing table and sends the packets to the alternate gateway. How can you be sure the routing table is correct and that it *has* the values entered that you need to support your internetwork? You can view the table by using the command line NETSTAT -NR or the ROUTE PRINT command. This command lists the IP addresses found in the table, the gateway that should be used when sending to each, and other information defining the nature and source of the routing information.

Depending on the destination host (which can be either a network or an individual workstation or server), the routing instructions can become elaborate. Several TCP/IP protocols are designed strictly for gateways to use when exchanging the routing information that enables them to ensure the most accurate and efficient routing possible—these include the Gateway-to-Gateway Protocol (GGP) and the Exterior Gateway Protocol (EGP). This prevents the traffic for each individual host from flooding the entire Internet in search of a single network.

The majority of TCP/IP routing is based on tables. Hosts and gateways each maintain their own routing tables and perform lookups of the destination addresses on incoming packets. While at one time the Internet relied on a collection of central core gateways as the ultimate source of routing information for the entire network, this has become impractical due to the tremendous growth of the Internet in recent years. Routing is now based on collections of autonomous systems called *routing domains* that share information with each other using the Border Gateway Protocol (BGP). Much of this is hidden to users of the Internet, and yet there are Web sites that show you the number of connections the datagrams make as they travel from the targeted host to your workstations and back again.

Getting to Know IP Multiplexing

After IP datagrams have been received by the host computer, they must be delivered to the transport layer protocol for interpretation and use by the targeted application service. On the sending workstation, multiplexing is the process of combining the requests made by several different applications into traffic for a few transport protocols, and then combining the transport protocol traffic into a single IP data stream. Multiplexing is really the process of taking several messages and sending them using a single signal. At the receiving computer, the process is reversed, in effect creating a demultiplexing routine of steps for in-bound messages.

The numeric values assigned to specific protocols and application services are defined on the host computer in text files named PROTOCOL and SERVICES, respectively. These are located in the Winnt/System32/drivers/etc folder in Windows 2000 Professional. Many of the values assigned to particular services are standardized numbers found in the Assigned Services RFC. FTP, for example, traditionally uses a port number of 20 on all types of host systems. Port numbers are individually defined for each transport protocol. In other words, both TCP and User Datagram Protocol (UDP) have different assignments for the same port number. The combination of an IP address and a port number is known as a socket.

INTEGRATING MICROSOFT TCP/IP INTO A NETWORK

Two market forces have led to the rapid adoption of TCP/IP into organizational networks: the growth of the Internet, and the need for integrating Windows NT and now Windows 2000 into heterogeneous networks. TCP/IP is now very much the foundation of connectivity of the Internet, and it is effectively leveling the playing field of operating system competition.

Many organizations today have multiple network operating systems running at the same time. A multitude of scenarios can lead to this condition, the most common of which being an integration of network architecture and protocols that are mainframe based, with both Novell NetWare and Windows 2000 being added to the total network as the number of users grows and their corresponding needs cause the first generation of network architecture to grow in order to meet evolving requirements. Another factor that drives organizations to have multiple network operating systems is the need users have for specific client/server applications that are supported on Windows 2000.

Organizations increasingly are finding that Novell NetWare has provided excellent file and print services, while Windows 2000, due to the pervasive support for Win32 application programmer interfaces (APIs) in the application development community, has the majority of 32-bit applications. Consequently, organizations have typically been integrating Windows NT as the application server. What's different about Windows 2000 Professional is that file and print services have been significantly improved, giving a Windows 2000-based dual processor workstation higher performance than its Novell-based counterpart networks.

Taken together, users' needs to run client/server applications that use data from UNIX servers and print services from NetWare creates a challenge for system administrators when it comes to troubleshooting problems. These separate protocols all cause redundant traffic and make it difficult to troubleshoot transmission problems over a network. Many administrators use TCP/IP as the unifying protocol in these kinds of scenarios, because TCP/IP is supported on all three platforms.

UNIX and its many variants, Windows NT, and the Internet all have TCP/IP protocol support at the kernel level, making interoperability possible. Novell's recent announcements on NetWare 5 showcase native TCP/IP support. However, the majority those organizations using NetWare today accomplish TCP/IP by using the NetWare/IP product. The bottom line is that TCP/IP is a protocol that spans hardware platforms, making it possible to integrate otherwise dissimilar network operating systems.

TCP/IP CONNECTIVITY UTILITIES

If you are using TCP/IP as the unifying protocol in a network of heterogeneous operating systems, a series of connectivity and diagnostic utilities are always useful to have for checking connections and ensuring reliability of the network. In fact, Windows NT contains its own versions of the standard TCP/IP utilities found on other platforms supporting this protocol. In global terms, TCP/IP utilities are separated into either connectivity-oriented utilities or diagnostic utilities used for monitoring a network and its clients.

The connectivity-oriented utilities are described here.

FINGER SERVICE

A command-line utility displays information about user(s) logged onto a remote system. The remote system must be running the Finger service for this command to function, and the output of this command varies depending on the remote system being addressed.

```
finger [-l] [username] @hostname
```

Table 3.1 describes each of the elements in this command.

FTP

Many times FTP is seen as a command, but it is in fact a protocol that uses the TCP/IP connection to transfer files to and from remote computers regardless of

TABLE 3.1 Parameters: Finger Service

Option	Description
-l	Displays information in long-list (verbose) format.
username	When this isn't defined, all users on the remote system are listed. Use this option to check the status of an individual user.
@hostname	Use this option for defining the IP address or hostname of the system you're querying.

the file systems being used at either end. In fact, FTP is the most widely used protocol within the TCP/IP suite because it is used for bridging diverse hardware platforms. This command originated on the UNIX platform and is used extensively in shell scripts for enabling communication between workstations and servers. Predominantly used for moving larger files between locations quickly and, for the most part, transparently to users, FTP is the glue that holds together client/server applications that use UNIX as a server component with Windows NT as client, and vice versa. As a command, FTP's goal is hardware interoperability between platforms.

You can use FTP both within shell scripts and interactively for moving files around a network or even the Internet. You can, in fact, use a web browser (Microsoft's Internet Explorer, for instance) to access FTP sites anywhere in the world providing that they are in the public domain and that you have the username and password to log in.

For a workstation to connect to another system via FTP, the destination system must be running an FTP server component. Microsoft delivers FTP as standard within Windows 2000 Professional. An FTP server for Windows 2000 Professional is delivered in Microsoft's Peer Web Services package, which began shipping in Windows NT 4.0 Workstation. You can configure a workstation based on Windows 2000 Professional as an FTP server using Microsoft Peer Web Services, making it possible to quickly post files and images for others to gather and use via the FTP command. In general, FTP is used from clients to larger file servers, where disk space is available that exceeds the capacities of a workstation.

FTP is also delivered as part of Microsoft's Internet Information Server (IIS) and ships with the Windows 2000 server operating system. In the instance of configuring IIS for FTP use, FTP actually runs as a network service that enables multiple users to connect with and use the server simultaneously. Nearly all UNIX operating systems run an FTP server daemon by default; FTP is often used for file transfers between UNIX and Windows NT systems. A daemon is the UNIX equivalent of a service under Windows NT. It is a program or utility than runs all of the time and provides resources that are available to any process that requests them.

Using the FTP command to connect with another system, you'll use the following syntax: When logging onto an FTP site, you'll be prompted to enter your username and password. If you are logging onto an anonymous FTP site, your username needs to be anonymous, and you use your e-mail address for the password. Almost all of the anonymous FTP sites do verification checking of e-mail addresses. After logging into the remote system, you can traverse the directories that have been made available by the system administrator through the definition of system security properties. If you're a system administrator, you'll find that Windows 2000's Peer Web Services can quickly configure an anonymous FTP site using the options in that service. Once you are logged onto a system using FTP, it is transparent to you whether the host system is based on Intel or Reduced Instruction Set Computing (RISC). The only way to tell which hardware platform you are logged onto is by looking at the subdirectory structure. FTP is indeed transparent to hardware platforms supporting it.

When you want to download files from an FTP server, you use the command GET. Conversely, loading files onto a server requires the command PUT. Some servers have restrictions on where files can be loaded, so be sure to check with the permissions set on your login and password before trying to load files to an FTP server. Keep in mind that the file names and commands themselves are case sensitive, which makes sense given the fact that the commands included in this protocol bridge both Windows NT (which is not case sensitive in its file structure) and UNIX (which is case sensitive in its definition of files). Compared to the performance from a 14.4- K or 28.8-K modem line, the FTP command is by far the most efficient method for downloading files from the Internet. Providing the organization from which you want to download files has an FTP site, you can get files of 2 megabytes (MB) or larger in a matter of seconds. Microsoft maintains a comprehensive FTP site at ftp.microsoft.com, which can be accessed through any Internet browser. Simply type the FTP location where you would normally type a web address, and the FTP site will be displayed. While the standard interface for the FTP command line is quickly becoming outdated because so many people are using browsers for traversing FTP sites, there are also utilities, both in the public domain and offered for sale that streamline the FTP process. Once FTP has been installed on Windows 2000 Professional on your system, you can use the command line below for accessing and downloading or updating files to selected FTP sites. If you're using a public-domain or purchased program for FTP functions, these commands are also used in those programs.

```
ftp [-v] [-n] [-i] [-d] [-g] [hostname] [-s:filename
```

TABLE 3.2 Parameters: FTP

Option	Description
-v	Prevents the display of remote server responses to client commands
-n	Prevents autologin upon connection
-i	Prevents individual file verifications during mass file transfers
-d	Displays debugging messages
-g	Allows wildcard characters to be used in file and directory names
hostname	Defines the host name or IP address of the remote system to be accessed
-s:filename	Allows you to specify a test file containing a series of FTP commands to be executed in sequence; in effect, launches a shell script of FTP commands to be executed on the remote system

Table 3.2 describes each of the elements of this command.

What about using the commands in the FTP protocol? Table 3.3 provides a quick tour of the most commonly used commands during FTP sessions.

TFTP

The Trivial File Transfer Protocol (TFTP) is based on using the UDP as the transport protocol; as a consequence, it is less secure and less reliable than its FTP counterpart, which uses TCP as its transport protocol. There are no user authentication services in TFTP. Further, TFTP does not provide for browsing of directories because the UDP protocol is not connection oriented. If you plan to use this command, you'll need to know the exact file name and location of what you need to retrieve.

```
tftp [ -i] host [get] [put] source {destination}
```

Table 3.4 describes each of the elements of this command.

TELNET

This is a terminal emulation program that makes it possible to log onto and use a remote workstation or server running telnet server services. Windows NT does not presently include a telnet server service. Many UNIX servers do have telnet server capabilities, and this command can be used from the Command Prompt window of a Windows 2000 Professional to log onto and use a UNIX-based server or any other server running a telnet service. The telnet implementation in Microsoft's TCP/IP supports DEC VT100, DEC VT52, or teletypewriter emulation (TTY) terminals via emulation.

Using the options associated with the Command Prompt window, you can define the terminal and display preferences for the telnet sessions you plan to initiate from a Windows 2000 Professional. Terminal emulation using telnet occurs through the Command Prompt window. Once connected with the

TABLE 3.3 Profile of Common FTP Commands

Command	Description
open *hostname*	Initiates a command session with a remote FTP host
close	Terminates the current session (without closing the FTP command window or program)
exit	Closes the FTP program, returning you to the command prompt of your system
ls	Lists all the files in the current directory
ls -l	Lists full information for the files in the current directory
cd */dirname*	Changes to the different directory
cd ..	Moves up one level in the directory tree
pwd	Displays the current directory name
binary	Sets the file transfer mode on FTP to binary or bit-by-bit mode (It's a good idea to use this command by default, especially on image-based files.)
ascii	Specifies that the file to be transferred is an ASCII file
get *filename*	Transfers the specified file to the local system
recv *filename*	Functions the same as GET; transfers the specified file to the local system
mget *filename*	Used in conjunction with wildcards in the command statement; transfers multiple files to the local system
put *filename*	Transfers the specified file to the remote system
send *filename*	Functions the same as put; transfers specified files to the remote system
mput *filename*	Transfers multiple files to the remote system; typically used in conjunction with wildcards
hash	Displays status of the current operation as files are transferred
prompt	Toggles the use of prompts for each individual file during multiple transfers
help	Provides the FTP command summary

TABLE 3.4 Parameters: TFTP

Option	Description
-i	Defines the file to be transferred in binary mode
host	Replaces the hostname or IP address of the remote system
get	Command for transferring a file or files from the remote system to the local system
put	Command for sending a file from the local system to the remote system
source	Defines the name of the file to be transferred
destination	Defines the location where the file is to be transferred

TABLE 3.5 Parameters: Telnet

Option	Description
host	Defines the hostname or IP address of the remote system you want to log onto.
port	Defines the port number in the remote system to which you will connect; when omitted, the value specified in the remote system's services file is used; if no value is defined in services, then port 23 is automatically used

remote system, the entire session can be completed from within the Command Prompt window. Telnet then works in conjunction with Windows 2000 using the emulations available via the Command Prompt window for communicating via the TCP protocol to other compatible systems.

```
telnet [host] [port]
```

TABLE 3.6 Parameters: RCP

Option	Description
-a	Specifies the file(s) to be transferred in ASCII mode
-b	Specifies the file(s) to be transferred in binary mode
-h	Allows files on a Windows NT system with file attributes set as hidden to be transferred
-r	Copies the contents of all of the source's subdirectories to the destination (when both source and destination are directories)
Source	Specifies the name of the file to be transferred (and optionally its host and usernames, in the format *host.user:filename*)
Destination	Specifies the name of the file to be created at the destination (and optionally its host and usernames, in the format *host.user:filename*)

Table 3.5 describes each of the elements of this command.

RCP

The purpose of this utility is to provide you with a command for copying files between a local system and a remote system that is running a remote shell server, rshd. Alternatively, you can direct two remote systems running rshd to exchange files between themselves using this protocol.

A text file on the remote system, called .RHOSTS, contains the host and user names of the local system so it is identified before the file transfers take place.

```
rcp [-a] [-b] [-h] [-r] source1 source2 … sourceN
destination
```

Table 3.6 describes each of the elements of this command.

REXEC

This utility provides for batched or noninteractive commands that are executed on a remote system. This command needs to have rexec service running on the remote system to be available to Windows NT clients. Redirection symbols can be used to refer output to files on the local system (using normal redirection syntax) or to files on the remote system (by enclosing the redirection symbols in quotation marks; for example, ">>").

```
rexec hostname [-l user] [-n] command
```

Table 3.7 describes each of the elements of this command.

TABLE 3.7 Parameters: Rexec

Option	Description
hostname	Specifies the hostname of the system on which the command is to be run
-l *user*	Specifies a username under whose account the command is to be executed on the remote system (a prompt for a user password will be generated at the local machine)
-n	Redirects rexec input to NUL
command	Specifies the command to be executed at the remote system

TABLE 3.8 Parameters: Rsh

Options	Description
hostname	Specifies the hostname of the system on which the command is to be run
-l *user*	Specifies a username under whose account the command is to be executed on the remote system (a prompt for a user password will be generated at the local machine)
-n	Redirects rsh input to NUL
command	Specifies the command to be executed at the remote system

Rsh

This command is used for executing commands and options on remote systems and requires that the rsh service be running on the remote system to be usable. It has the same redirection capabilities as rexec and the same .rhosts requirement as RCP.

```
rsh hostname [-l user] [-n] command
```

Table 3.8 describes each of the elements of this command.

LPR

This utility provides the flexibility of printing a file to a printer connected to a remote Berkeley System Distribution(BSD)-type printer subsystem that is running an LPR server.

```
lpr -Sserver -Pprinter [-Jjobname] [-ol] filename
```

Table 3.9 describes each of the elements of this command.

TABLE 3.9 Parameters: LPR

Options	Description
-S*server*	Specifies the hostname of the system to which the printer is connected
-P*printer*	Specifies the name of the printer to be used
_J*jobname*	Specifies the name of the print job
-ol	Used when printing a nontext (PostScript) file from a Windows NT system to a UNIX printer
-l	Used when printing a nontext (PostScript) file from a UNIX system to a Windows NT printer
filename	Specifies the name of the file to be printed

THE DYNAMIC HOST CONFIGURATION PROTOCOL

There's a fundamental difference between static IP addresses and those assigned on an as-needed basis. In the latter case, the role of the DHCP protocol is necessary for quickly getting IP addresses assigned when a given user needs them, as is the case with many ISPs. In addition, the role of DHCP in companies is greatly expanding as more and more employees are dialing in to get their e-mail and check websites. The ability of DHCP to dynamically assign a different IP address also has a positive security aspect to it as well; when a PC has a different IP address every time, it's more difficult to break into over the Internet.

What Is DHCP?

Let's say you are responsible for supporting 50 to 100 people using laptops who travel extensively and then come back into the office to check email and download files for use at customer sites. How can you create a network that makes it possible for members of this group to get full TCP/IP access while in the office or even using Remote Access Services (RAS) from regional offices equipped with fractional T1 lines? Instead of choosing the NetBEUI protocol and settling for its limitations from both a performance and security perspective, how can you enable this group with TCP/IP access? Through the use of DHCP originally developed Microsoft. This protocol truly functions just like a library that checks books in and out, only this protocol checks IP addresses in and out. Best of all, this IP address mechanism is for the most part administered at the server level, alleviating ongoing configuration at the client level of TCP/IP configuration options. This means that, once a laptop is configured with the ability to accept an IP address, the user at the remote location doesn't need to configure any other parameters.

DHCP is an element of the TCP/IP protocol suite that enables you as a system administrator to automatically configure parameters such as IP addresses, subnet masks, and default gateways for remote systems that have sporadic access to a TCP/IP-based network. A DHCP server running a Windows 2000 server manages these attributes responsible for handling a TCP/IP connection. Microsoft, working in conjunction with other companies, created this protocol to solve the dilemma of how to more easily assign IP addresses to workstations, large networks, and mobile users who need IP addresses when they dial into a network.

The DHCP server is included in the baseline of Windows NT Server 4.0 and the Windows 2000 Server. The DHCP server consists of two components: a mechanism for tracking and allocating TCP/IP configuration parameters, and a protocol that can distinguish when an IP address has been delivered to and acknowledged by the client. The DHCP standard, published in the Internet Engineering Task Force's RFC 1541, has been strongly influenced by BOOTP protocol defined in RFC 951. BOOTP is a segment of the TCP/

IP command set that was originally developed for use with diskless terminals needing TCP/IP compatibility.

BOOTP was first developed for these diskless workstations from the server side to ensure that TCP/IP connections could be made and the TCP/IP configuration settings delivered. While BOOTP is capable of assigning the same IP address to a diskless workstation or terminal, DHCP can dynamically assign and renew workstation configurations from a pool of IP addressees, solving many of the problems that have hampered the ongoing acceptance of the BOOTP protocol. The Windows NT server ships with integrated DHCP server modules that are compatible with the TCP/IP command set, regardless of what operating system the originating system uses.

WHY DHCP WAS CREATED?

With the widespread growth of large-scale networks, the integration of multiple needs within large organizations, and the need for uniquely identifying systems on a large, packet-switched network as the Internet, it became important to be able to selectively identify each client quickly. The identification of client systems needed to be transparent to the hardware; the approach had to be based solely on stable protocols that could traverse the hardware differences between systems.

Different types of networks have varying approaches to identifying and communicating with systems that are part of their topology. An Ethernet or Token Ring LAN uses a unique MAC address hard-coded into each NIC by the NIC's manufacturers. Because each manufacturer numbers its cards sequentially, and because part of that MAC address consists of a code identifying the manufacturer itself, the device's address is unique not only on the local network where it is used, but on all networks everywhere. No other adapter exists that uses the exact same address.

This hard-coded approach would be a pervasive solution if all users of the Internet had a compatible system that could communicate these protocols. For many networks, system administrators actually override the hardware address with their specific IP addressing approach. For the majority of Internet users today, there are simply no NIC cards installed in their systems—they are using modems to dial up their ISPs to gain access to the Internet. This approach to connectivity to the outside world necessitated a network protocol that could effectively assign IP addresses dynamically from a pool or inventory of addresses, ensuring that the dial-up system has a unique identity.

UNDERSTANDING HOW DHCP WORKS

When a client is configured to use DHCP, it acquires its IP configuration settings during the workstation's initialization and booting process. The best way to think of DHCP working is to visiualize books being checked out of a library: A book is the IP address being assigned to a given PC at the time of login to the network. This process relies heavily on the BOOTP protocol to initialize the entire process of finding the DHCP server on a network. DHCP runs on top of UDP for communications, with clients transmitting to the DHCP server and services using the RFC assigned numbers. Windows 2000's DHCP communications are defined through the use of the RPC application programming interface. All DHCP communications use the same packet format.

Let's walk through DHCP's major steps to see how it works.

- First, a DHCP client broadcasts to all DHCP servers on the network, using the BOOTP protocol to transmit its request.

- It is the responsibility of the routers and computers functioning as DHCP relay agents on the local network to propagate or forward the traffic generated by a client to other networks where additional DHCP servers may be located.

- If the workstation looking for an IP address using DHCP does not find one, it will revert back the settings for the previous IP address that were assigned in a previous session.

- When a client system does get an IP address assigned to it, it is called an *IP lease*, or the act of leasing an address.

If another client is actually using the options defined in the previous settings, an error message will be sent back to the client. Windows 2000 Professional will in this instance repeat its attempt to contact a DHCP server until an appropriate response is received, generating an error each time the predetermined time-out values are reached.

A workstation configured as a DHCP client that has no previous configuration settings assigned to it by a server must begin the client configuration sequence from the beginning. As an administrator it's important that you realize the circumstances that can leave a DHCP client in an unconfigured state. Here are the specific examples of how a workstation running Windows 2000 Professional configured as a DHCP client can encounter unconfigured states of use: The workstation has configuration options as defined from a DHCP client session whose lease has expired.

- The workstation has just been configured and has yet to receive a DHCP client setting parameters options.

- The workstation has moved from its existing subnet to a new subnet, which is unrecognized by the DHCP server that originally defined the IP lease.

- The workstation may have explicitly released its hold on previous IP addressing options.

INSTALLING DHCP CLIENT SERVICES

This protocol is designed to alleviate the complexities of configuring TCP/IP addresses and to give client workstations the flexibility of having addresses assigned to them dynamically. This is particularly valuable when a laptop is running Windows NT Workstation and needs access to either an intranet within a company or to the Internet via an ISP. After installing TCP/IP and, with it, support for DHCP, every time Windows 2000 Professional is rebooted, the TCP/IP stack is reinitialized and the DHCP parameters are again communicated onto the network.

Here are the steps for configuring DHCP in Windows 2000 Professional:

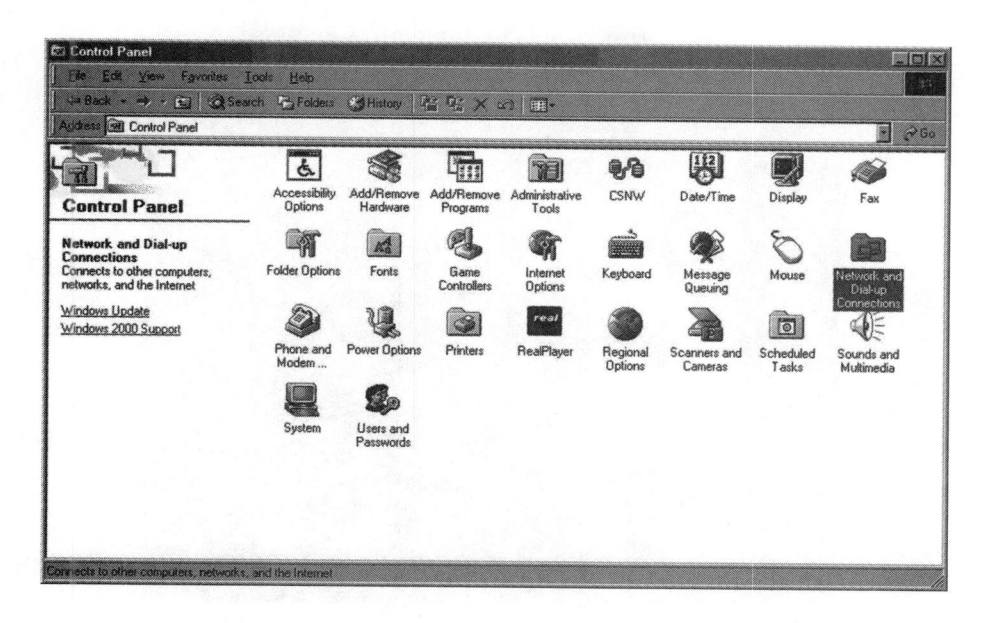

CHAPTER 3: INTRODUCTION TO TCP/IP AND NETWORKING

FIGURE 3.2 The Network Utility in the Control Panel

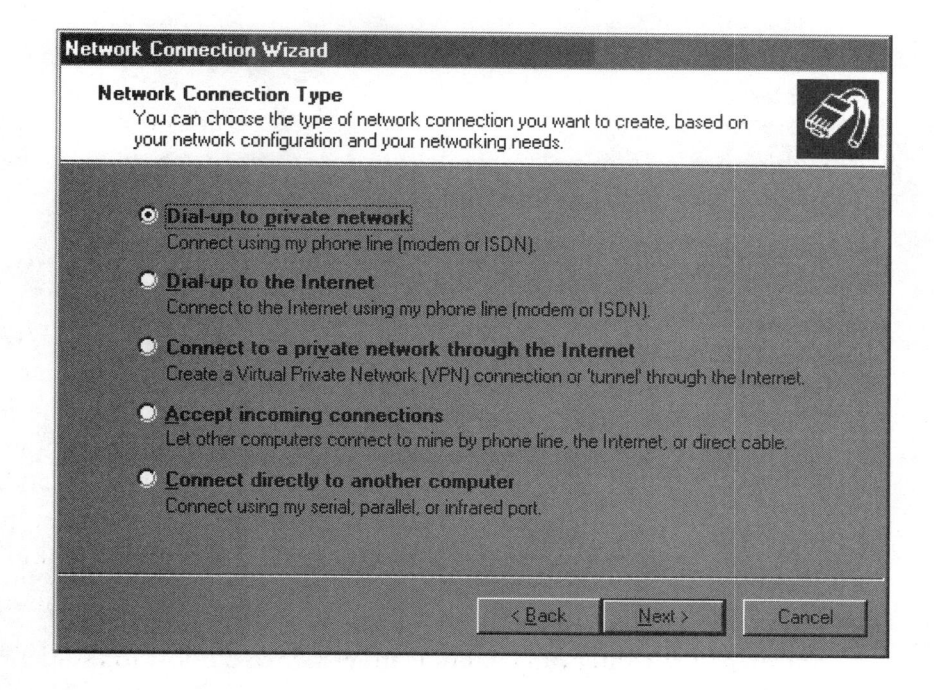

FIGURE 3.3 The Options in the Network Properties Dialog Box

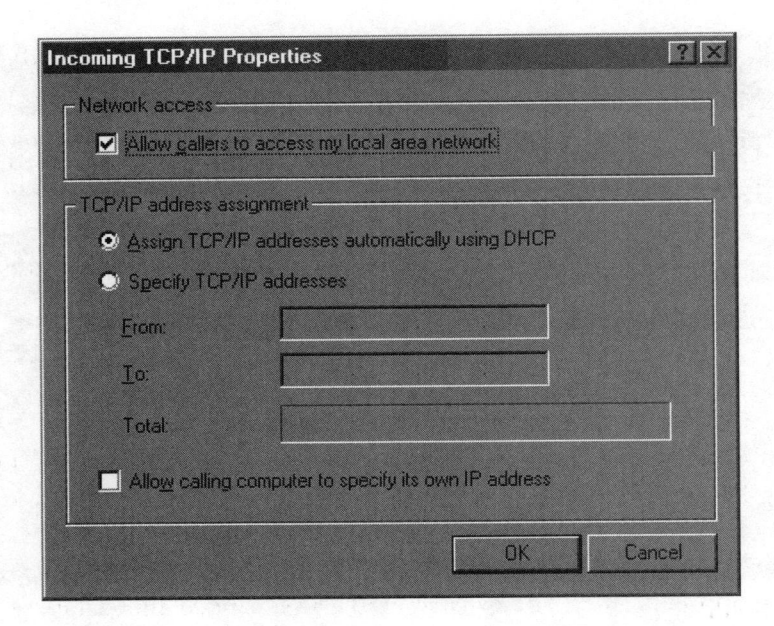

FIGURE 3.4 Using the Protocols Page to Configure DHCP

1. Double-click on My Computer, located in the upper left corner of the Desktop.

2. Double-click on Control Panel. Figure 3.2 shows the contents of the Control Panel with the Network icon highlighted.

3. Double-click on the Network icon. The Network Wizard appears(see Figure 3.3).

   Click once on the Protocols page (see Figure 3.4).

4. Click once on the TCP/IP Protocol entry on the Protocols page of the Network Properties dialog box.

5. Click once on the Incoming TCP/IP Properties page (see Figure 3.4). Use the options in this dialog box to select the DHCP protocol as the one you'll be using in conjunction with this workstation. Be sure to delete any of the static IP addresses defined in previous configuration sessions.

6. Click once on Apply. The changes you have defined are then applied to your system.

7. Click once on OK. The modifications to your system are then made. Follow the options defined in the series of dialog boxes presented to reboot the system. Once the workstation has been rebooted, DHCP is activated and seeks out a DHCP server to gain an IP address.

USING DNS IN WINDOWS 2000

As with many of the advances made in the TCP/IP protocol, Domain Naming Services (DNS) was created during the 1980s to accommodate the growth of the Internet and the need for a reliable naming service. Just as DHCP was specifically developed to dynamically assign IP addresses through a leasing approach, DNS is a distributed name service and database system that spreads the load of system identification around the network by dividing the it into domains. Domain names are then are registered throughout the network rather than individual systems or workstations. The unique name of the user and the domain it is a member of is the basis of the DNS databases that are distributed throughout networks based on this approach.

A DNS server is like a WINS server in that both are databases of computer names (in this case, IP host names), but the comparison stops there. Compared to WINS, a DNS server by itself is somewhat limited in its ability to parse names and domains. A DNS host does not have the ability to handle automatic name registration and must be manually updated each time a change is made to the network.

How DNS Naming Resolution Happens

DNS servers are actually centrally located host tables, where the names of individual workstations are compared and then cross-referenced for the originating system, making it possible to communicate with a targeted server. Requests for the identity of a workstation or node are sent to a DNS server, which then looks into the DNS database and does or does not return the IP address corresponding to the targeted workstation. Its underlying strength is evident only when users try to resolve the identities (addresses) of computers located at remote Internet sites. In the distributed architecture of DNS, there is no single listing or database that contains the addresses of all the computers using the Internet. There is instead a collection of computers known as root servers that contain a complete listing of the domain names registered to individual networks. The entire DNS architecture is heavily based on a tree-like structure, emanating out of the root servers. The listings of root servers for each domain include the IP addresses of the DNS machines that have been designed as the defining servers for DNS-based traffic.

Despite its limitations, the DNS architecture makes it possible for administrators to easily manage the network traffic specific to and relying on the DNS servers in Windows NT-based domains. Every DNS server uses its host tables as pointers for messages being sent from initiating client workstation to destinations. By beginning at the root of the Internet tree, you can locate any host on any computer by parsing the database from its root domain down to its hostname and consulting the databases for each level.

The Role of the HOSTS File

In the context of DNS, the HOSTS file is the deliverable or the item where the host or workstation name and IP address are recorded. Since the inception of the TCP/IP protocol, the HOSTS file has actually been the clearinghouse where resolution of name and IP address is completed. The HOSTS file was challenged in its performance by the rapid adoption of TCP/IP networking and, with it, the triple-digit compound annual growth rates of systems of all types needing IP addresses. However, the HOSTS file, in smaller configuration, does provide an efficient name resolution approach because it is consulted before communications begins.

The Windows 2000 Professional and server versions of the HOSTS file is actually called LMOSTS (LAN Manager HOSTS file) and is used for NetBIOS name resolution. If your organization is connecting with only a few Internet hosts, you can increase the connection efficiency by adding them to the workstation's LMHOSTS file. Keep in mind that performance is maximized when there are up to 15 website addresses in the LMHOSTS file. Beyond 15, performance tends to get bogged down. The IP addresses in the LMHOSTS file are also to be listed according to priority—the most commonly used IP addresses need to be at the top of the file for the best performance.

SUMMARY

Considering the growth of the Internet, and with it the explosive growth in the number of workstations of all types using the Internet, there is a continual need to evaluate and expand the capabilities of the network protocols used to assign names and IP addresses to these systems. The intent of this chapter has been to provide you with an overview of the single biggest connectivity force in the world of operating systems and connectivity today: the TCP/IP networking protocol. TCP/IP has served to level the playing field of operating systems and workstations of all types by making connectivity a given. No longer are workstations connected via proprietary operating systems, since Windows NT's tight integration of TCP/IP has made workstations based on Windows 2000 Professional capable of communicating with virtually any other system that is compatible with the TCP/IP protocol. The integration of naming conventions into the TCP/IP standard continues to improve the performance of these networks. DHCP streamlines naming by providing a lease-based approach to defining IP addresses, and DNS provides a database-driven approach to alleviating naming conflicts.

CHAPTER 4

Introducing Home and Small Business

LAN CONNECTIVITY BASICS

Network connectivity products are now under $50. Of all the high-speed LAN alternatives, Ethernet is the least expensive. Ethernet adapter cards transmit and receive data at speeds of 10 mbps through up to 300 feet of telephone wire to a hub device normally stacked in a wiring closet. The hub adds a minimal cost to each desktop connection. Data is transferred between wiring closets using either a heavy coax cable (Thicknet) or fiber-optic cable.

Most textbook treatments of Ethernet have concentrated on Thicknet coax because that is the wiring arrangement used when Xerox invented the LAN. Today it is still used for medium-to-long distances where medium levels of reliability are needed. Fiber goes farther and has greater reliability, but at a higher cost. To connect a number of workstations in the same room, a light-duty coax cable (Thinnet) is commonly used. These media reflect an older view of workstation computers in a laboratory environment.

However, the PC and Macintosh have changed the geography of networking. Computers are now located on desktops, in dorm rooms, and at home. Telephone wire is the clear choice (where possible) for the last hop from basement to desktop.

Drivers to support the PC Ethernet card come in four versions:

Access to the Internet under DOS can be provided using one of the packet driver programs. A collection of free drivers is available from various Internet servers.

Support for Novell clients under DOS can be packaged as a module called IPX.COM.

When Novell must share Ethernet with other software, it supplies a proprietary interface called Output Device Interface (ODI). Because of the large market share controlled by Novell, ODI supports most adapter cards and is used by several other software vendors.

All major companies (e.g., Microsoft, IBM, DEC, AT&T) and operating systems (e.g., Windows for Workgroups, OS/2, NT, 2000, and Windows XP) use Network Device Interface Specification (NDIS). Developed jointly by Microsoft and 3Com, NDIS also supports most adapter cards and is the native choice for Windows and OS/2 peer networks.

Through NDIS or ODI it is possible to support Novell IPX, IBM SNA, DECNET, Appletalk, TCP/IP (for the Internet), and NetBIOS, all simultaneously. Of course, it takes a very large machine and an advanced operating system to squeeze all this software into memory.

This chapter will explain the basic elements of Ethernet to a PC user. It assumes that someone else will probably be purchasing the central equipment and installing the wire.

DEFINITIONS AND STANDARDS

The early development of Ethernet was done by Xerox Research. Its name was a registered trademark of Xerox Corporation. The technology was refined, and a second generation called Ethernet II was widely used. Ethernet from this period is often called DIX after its corporate sponsors Digital, Intel, and Xerox. As the holder of the trademark, Xerox established and published the standards.

Obviously, no technology could become an international standard for all sorts of equipment if the rules were controlled by a single U.S. corporation.

The IEEE was assigned the task of developing formal international standards for all LAN technology. It formed the 802 committee to look at Ethernet, Token Ring, Fiber Optic, and other LAN technology. The objective of the project was not just to standardize each LAN individually, but also to establish rules that would be global to all types of LANs so that data could easily move from Ethernet to Token Ring or Fiber Optic.

This larger view created conflicts with the existing practice under the old Xerox DIX system. The IEEE was careful to separate the new and old rules. It recognized that there would be a period when old DIX messages and new IEEE 802 messages would have to coexist on the same LAN. It published a set of standards of which the most important are these:

- 802.3—Hardware standards for Ethernet cards and cables
- 802.5—Hardware standards for Token Ring cards and cables
- 802.2—The new message format for data on any LAN

The 802.3 standard further refined the electrical connection to Ethernet and was immediately adopted by all the hardware vendors. Today all cards and other devices conform to this standard.

The 802.2 standard, however, would require a change to the network architecture of all existing Ethernet users. Apple had to change its Ethertalk, and did so when converting from Phase 1 to Phase 2 Appletalk. DEC had to change its DECNET. Novell added 802 as an option to its IPX, but it supports both DIX and 802 message formats at the same time.

The Internet Engineering Task Force (IETF) group, which manages Internet standards, refused to change the TCP/IP protocol used by the Internet, and decided to stick with the old DIX message format indefinitely. This produced a deadlock between two standards organizations that has not been resolved.

IBM waited until the 802 committee released its standards and then rigorously implemented the 802 rules for everything *except* TCP/IP where the IETF rules take precedence. This means that NetBEUI (the format for NetBIOS on LAN) and System Network Architecture (SNA) obey the 802 conventions.

So Ethernet suffers from too many standards. The old DIX rules for message format persist for some uses (Internet, DECNET, some Novell). The new 802 rules apply to other traffic (SNA, NetBEUI). The most pressing problem is making sure that Novell clients and servers are configured to use the same frame format.

ACCESS AND COLLISIONS

Ethernet uses a protocol called CSMA/CD, which stands for Carrier Sense, Multiple Access, Collision Detect. The *Multiple Access* part means that every station is connected to a single copper wire (or a set of wires that are connected together to form a single data path). The *Carrier Sense* part indicates that, before transmitting data, a station checks the wire to see if any other station is already sending something. If the LAN appears to be idle, then the station can begin to send data.

An Ethernet station sends data at a rate of 10 Mbps. A bit allows 100 nanoseconds per bit. Light and electricity travel about 1 foot in a nanosecond. Therefore, after the electric signal for the first bit has traveled about 100 feet down the wire, the station begins to send the second bit. However, an Ethernet cable can run for hundreds of feet. If two stations happen to be located 250 feet apart on the same cable and both begin transmitting at the same time, they will then be in the middle of the third bit before the signal from each station reaches the other station.

This explains the need for the *Collision Detect* part. Two stations can begin to send data at the same time, and their signals will collide nanoseconds later. When such a collision occurs, the two stations stop transmitting, "back off," and try again later after a randomly chosen delay period.

While an Ethernet LAN can be built using one common signal wire, such an arrangement is not flexible enough to wire most buildings. Unlike an ordinary telephone circuit, Ethernet wire cannot simply be spliced together, connecting one copper wire to another. Ethernet requires a repeater, which is a simple station connected to two wires. Any data that the station receives on one wire it repeats bit for bit on the other wire. When collisions occur, the station repeats the collision as well.

In common practice, repeaters are used to convert the Ethernet signal from one type of wire to another. In particular, when the connection to the desktop uses ordinary telephone wire, the hub back in the telephone closet contains a repeater for every phone circuit. Data coming down any phone line is copied onto the main Ethernet coax cable, and data from the main cable is duplicated and transmitted down every phone line. The repeaters in the hub electrically isolate each phone circuit, which is necessary if a 10-Mb signal is going to be carried 300 feet on ordinary wire.

Each set of rules is best understood by characterizing its worst case. The worst case for Ethernet starts when a PC at the extreme end of one wire begins sending data. The electric signal passes down the wire through repeaters, and, just before it gets to the last station at the other end of the LAN, that station (hearing nothing and thinking that the LAN is idle) begins to transmit its own data. A collision occurs. The second station recognizes this immediately, but the first station will not detect it until the collision signal retraces the first path all the way back through the LAN to its starting point.

Any system based on collision detection must control the time required for the worst round-trip through the LAN. As the term *Ethernet* is commonly defined, this round-trip is limited to 50 microseconds (millionths of a second). At a signaling speed of 10 Mbps, that is enough time to transmit 500 bits. At 8 bits per byte, this is slightly less than 64 bytes.

To make sure that the collision is recognized, Ethernet requires that a station continue transmitting until the 50-microsecond period has ended. If the station has less than 64 bytes of data to send, then it must pad the data by adding zeros at the end.

In simpler days, when Ethernet was dominated by heavy-duty (Thicknet) coax cable, it was possible to translate the 50-millisecond limit and other electrical restrictions into rules about cable length, number of stations, and number of repeaters. However, by adding new media (such as fiber-optic cable) and smarter electronics, it becomes difficult to state physical distance limits with precision. However those limits work out, they are ultimately reflections of the constraint on the worst-case round-trip.

It would be possible, however, to define some other Ethernet-like collision system with a 40- or 60-microsecond period. Changing the period, the speed, and the minimum message size simply requires a new standard and some alternate equipment. AT&T, for example, once promoted a system called Starlan that transmitted data at 1 Mbps over older phone wire. Many such systems are possible, but the term *Ethernet* is generally reserved for a system that transmits 10 Mbps with a round-trip delay of 50 microseconds.

A repeater receives and then immediately retransmits each bit. It has no memory and does not depend on any particular protocol. It duplicates everything, including the collisions. To extend the LAN farther than the 50-microsecond limit will permit, you need a bridge or router. These terms are often confused:

A *bridge* receives the entire message into memory. If the message was damaged by a collision or by noise, then it is discarded. If the bridge knows that the message is being sent between two stations on the same cable, then it discards it. Otherwise, the message is queued up and to be retransmitted on another Ethernet cable. The bridge has no address. Its actions are transparent to the client and server workstations.

A *router* acts as an agent to receive and forward messages. The router has an address and is known to the client or server machines. Typically, machines send messages to each other directly when they are on the same cable, and they send messages addressed to another zone, department, or subnetwork to the router. Routing is a function specific to each protocol. For IPX, the Novell server can act as a router. For SNA, an APPN Network Node does the routing. Dedicated devices, UNIX workstations, or OS/2 servers do the routing for TCP/IP.

PROBLEM DETERMINATION

In the classical fat yellow Ethernet cable, Thicknet, a heavy copper wire is embedded in plastic and surrounded by a grounded metal shield. At each end of the cable, the central signal wire connects to a resistor that in turn connects to the grounded shield. As previously noted, each bit follows the previous bit at a distance of about 100 feet. The bits are represented by a wave of electrical voltage. The resistor at each end of the wire removes the signal cleanly from the wire. Without such a termination, some part of the voltage wave would hit the end of the wire and bounce back, causing confusion and perhaps appearing as a collision.

Ethernet LANs fail in three common ways. A nail can be driven into the cable breaking the signal wire. A nail can be driven in touching the signal wire and shorting it to the external grounded metal shield. Finally, a station on the LAN can break down and start to generate a continuous stream of junk, thereby blocking everyone else from sending.

There is a specialized device that finds problems in an Ethernet LAN. It plugs into any attachment point in the cable and sends out its own voltage pulse. The effect is similar to a sonar "ping." If the cable is broken, then there is no proper terminating resistor. The pulse will hit the loose end of the broken cable and bounce back. The test device senses the echo, computes how long the round-trip took, and then reports how far away the break in the cable is.

If the Ethernet cable is shorted out, a simple voltmeter can determine that the proper resistor is missing from the signal and shield wires; again, by sending out a pulse and timing the return, it can determine the distance to the problem. Most of the thinking about Ethernet repair has been based on the original Thicknet media. However, modern Ethernet installation may not use this old coax cable. The connection to the desktop may be based on telephone wire between the PC and a hub device. The hubs may stack up in a wiring closet and then be connected to other rooms using fiber-optic cable.

Newer generations of *smart* hubs can perform part of the error detection and reporting function. For example, they could isolate a problem in the connection to a particular desktop workstation and automatically isolate that unit from the rest of the network.

Ethernet presents a classic trade-off. The simplest equipment has a very low cost, but requires some technical expertise to locate and repair errors. More sophisticated equipment may be able to do automatic error detection and recovery, but costs more to buy.

FRAME FORMATS

A block of data transmitted on Ethernet is called a *frame*. The first 12 bytes of every frame contain the 6-byte destination address (the recipient) and a 6-byte source address (the sender). Each Ethernet adapter card comes with a unique factory-installed address (the *universally administered address*). Use of this hardware address guarantees a unique identity to each card.

The PC software (in PROTOCOL.INI or NET.CFG) can be configured to substitute a different address number. When this option is used, it is called a *locally administered address*. If the use of this feature is properly controlled, the address can contain information about the building, department, room, machine, wiring circuit, or owner's telephone number. When accurate, such information can speed problem determination.

The source address field of each frame must contain the unique address (universal or local) assigned to the sending card. The destination field can contain a *multicast* address representing a group of workstations with some common characteristic. A Novell client might broadcast a request to identify all NetWare servers on the LAN, while a Microsoft or IBM client machine might broadcast a query to all machines supporting NetBIOS to find a particular server or domain.

In normal operation, an Ethernet adapter will receive only those frames with a destination address that matches its unique address or the frames with destination addresses that represent a multicast message. However, most Ethernet adapters can be set into *promiscuous* mode; when this happens, they receive all frames that appear on the LAN. If this poses a security problem, a new generation of smart hub devices can filter out all frames with private destination addresses belonging to another station.

There are three common conventions for the format of the remainder of the frame:

1. Ethernet II or DIX
2. IEEE 802.3 and 802.2
3. snap

ETHERNET II OR DIX

Before the development of international standards, Xerox administered the Ethernet conventions. As each vendor developed a protocol, a 2-byte-type code was assigned by Xerox to identify it. Codes were given out to XNS (Xerox's own protocol), DECNET, IP, and Novell IPX. Because short Ethernet frames must be padded with zeros to make them 64 bytes in length, each of these higher-level protocols required either a larger minimum message size or an internal length field that could be used to distinguish data from padding.

Type field values of particular note include:

0%0600 XNS (Xerox)

0%0800 IP (the Internet protocol)

0%6003 DECNET

IEEE 802.3 AND 802.2

The IEEE 802 committee was charged to develop protocols that could operate the same way across all LAN media. To allow for collision detect, the 10-Mb Ethernet requires a minimum packet size of 64 bytes. Any messages shorter than that must be padded with zeros. The requirement to pad messages is unique to Ethernet and does not apply to any other LAN media. In order for Ethernet to be interchangeable with other types of LANs, it would have to provide a length field to distinguish significant data from padding.

The DIX standard did not need a length field because the vendor protocols that used it (XNS, DECNET, IPX, IP) all had their own length fields. However, the 802 committee needed a standard that did not depend on the good behavior of other programs. The 802.3 standard therefore replaced the 2-byte type field with a 2-byte length field.

Xerox did not assign any important types to have a decimal value below 1,500. Because the maximum size of a packet on Ethernet is 1,500 bytes, there was no conflict or overlap between DIX and 802 standards. Any Ethernet packet with a type/length field of less than 1,500 is in 802.3 format (with a length field), while any packet in with a field value greater than 1,500 must be in DIX format (with a type field).

The 802 committee then created a new field to substitute for type. The 802.2 header follows the 802.3 header (and also follows the comparable fields in a Token Ring, Fiber Distributed Data Interface (FDDI), or other types of LAN).

The 802.2 header is 3 bytes long for control packets or the kind of connectionless data sent by all the old DIX protocols. A 4-byte header is defined for connection-oriented data, which refers primarily to SNA and NetBEUI. The first 2 bytes identify the Service Access Point (SAP). Even with hindsight it is not clear exactly what the IEEE expected this field to be used for. In current use, the two SAP fields are set to 0×0404 for SNA and 0×F0F0 for NetBEUI.

SNAP

The IEEE left all the other protocols in a confusing situation. They did not need any new services and did not benefit from the change. Furthermore, a 1-byte SAP could not substitute for the 2-byte type field. Yet 802.2 was an international standard, and that has the force of law in many areas. The compromise was to create a special version of the 802.2 header that conformed to the standard but actually repackaged the old DIX conventions.

The first 5 bytes of what 802.2 considers data are actually a subheader ending in the 2-byte DIX type value. Any of the old DIX protocols can convert their existing logic to legal 802 snap by simply moving the DIX type field back 8 bytes from its original location.

SUMMARY

The textbook definitions of Ethernet have little to do with current practice. Ethernet is supposed to be a single common medium with multiple connections. That may be true for older installations and laboratory environments; however, new desktop installations bring Ethernet to the desktop over phone wire and frequently build the spine using fiber optics.

The connection between the hub in the wiring closet and the adapter card in the PC forms a single point-to-point Ethernet segment between two stations. The connection to the rest of the LAN involves active electronics in the hub. In current use, this is done with a repeater that copies every bit and propagates collisions.

A new generation of even smarter hubs provides a bridge connection between the main LAN and the phone wire. Only multicast messages and private messages specifically addressed to the PC are forwarded to the desktop. This has two advantages:

- It provides greater security, because the desktop user cannot spy on traffic addressed to other nodes.

- It provides each desktop user with an isolated, private 10-Mb data path free of collisions.

- The connection between hubs can then use a higher-speed fiber-optic protocol (such as Asynchronous Transfer Mode [ATM]) to deliver much greater performance than simple Ethernet. This hybrid (Ethernet to the desktop, something else between the hubs) represents a compromise of high performance and low cost.

However, bridging Ethernet to any other LAN protocol requires some attention to frame formats. Unfortunately, the standards are still a mess. DIX and 802 messages flow on the same LAN. Bridges must be aware of the protocol conventions and select the correct frame format when moving data onto or off of an Ethernet LAN.

CHAPTER 5

MANAGING SERIAL DEVICES IN A NETWORKED ENVIRONMENT

The proliferation of network management technology in both hardware and software has provided the network manager and now the home computer user with an extensive tool kit. Still, many devices being used today support only serial management. Such devices provide a serial management port for attaching terminals or PCs to perform management functions. These serial-only devices may also require the network manager to implement costly and sometimes cumbersome workarounds to be included in the management scheme. This chapter explores the technological advances in Device Server design that make it possible to manage serial-only devices over a network.

MANAGEMENT VIA SERIAL COMMUNICATIONS

Nonnetworked Techniques

To current PC users, networked communication seems like a birthright. However, remember that for many years serial communication was the only way to transmit data from one device to another. Many industrial, medical, and monitoring devices were designed long before the current advances in integrated circuits made Ethernet technology so inexpensive. Some of these devices would even qualify as legacy equipment, built with internal processing capabilities far below that of current desktop PCs.

The only management option provided on many of these products is a serial port, usually RS-232. This port transmits data bits one after another (hence the name "serial" port). Many times a PC will act as a "server" for the serially streamed data, completing packet creation and organization of the data before it is transmitted over the network to another PC or peripheral.

Typically, these devices can be managed remotely only by adding external devices allowing access to the dedicated management PC or by attaching directly to the serial port itself.

In the past, remote management of devices with serial ports was accomplished by running a long serial connection using RS-232 cables. For greater distances, either RS-422 cables were used if the device itself supported this standard, or some type of external converter was installed.

Networked Techniques

In the 1980s and 1990s, terminal servers became a popular way to connect serial devices to a network for accessing multiuser host systems. In a terminal server design, serial devices were physically connected to terminal server ports. The terminal server itself governed protocol use and the availability of added features such as multiple connection support from a single port, or dedicated connections to a single system on power-up. TCP/IP was the protocol suite of choice for this type of configuration because it allowed more efficient communications over both the local network (intranet) and the Internet.

The primary function of terminal servers was to give terminal users access to network devices. Connecting network devices to remote serial ports became another popular use. Printers, modems, converters, and other task-specific devices could be attached to server ports, and these ports could then be accessed from network hosts. In this configuration, monitoring devices could be polled at regular intervals; serial console ports on computers could be accessed from remote locations for important boot-time parameter configuration; finally, the server itself could be accessed to examine performance data collected on the various ports themselves. While some users simply used IP's Telnet for management connections, other management-specific suites, such as Simple Network Messaging Protocol (SNMP), were developed to further enhance status and configuration management of these remote device ports.

DEVICE SERVER TECHNOLOGY

Whereas earlier terminal server products were somewhat large, requiring substantial processing resources, newer integrated circuit designs have made it possible shrink the size of a terminal server. As we begin the new millennium, new single-port Device Servers can be built that are cost effective and small in size. The availability of single-port device servers created the potential for isolated single-port serial devices to be networked in a cost-effective manner.

One can see very quickly how a network manager can now attach the serial port of a device to a single-port server and immediately have network access to that device. No long cable runs, dedicated modem, or multiplexer ports are now required—simply install the device server on the nearby network, attach it to the serial port of the device, and manage the device from anywhere within the corporate or campus network (intranet) or from the Internet.

Some devices require a dedicated PC not only to present the management and configuration information, but also to process it before the user views it. Can this type of device also benefit from the availability of a device server? The answer is yes.

By using a redirector software package, a PC running software specifically to process information from a serial device can become a networked PC. The redirector takes the PC's output destined for its communications port and redirects it to a network port—specifically, to the network port on the device server. Thus, the PC "believes" it is talking to a local device on the COM port when it is really talking to a device located remotely on the network.

Even without a redirector, a remote PC can be connected via a process called *tunneling*. Here, each device server passes the serial data from one end of the connection to the other. This configuration can be utilized if the data between the serial device and the PC is encoded or proprietary.

THE BENEFITS OF NETWORKED MANAGEMENT

Device server technology enables an isolated device to be networked into the campus or corporate network. Why network these devices? Several reasons come to mind:

1. Easy Installation and Maintenance. Network connections tend to populate every location of a campus or corporate site. Wherever one goes, a network access port is usually nearby. This means a device in any location can be put onto the network and accessed from anywhere else on the local network or even over the Internet. As networks are extended to great lengths using switches, hubs, and converters, connectivity becomes available to areas that previously required long dedicated serial cable runs.

2. Management from Anywhere. Network managers now have a great many tools at their disposal for ensuring that the network performs efficiently. SNMP (including management information bases [MIBs])is a standardized management protocol providing proactive management information arising from continuous process monitoring. Many vendors, such as Hewlett Packard (HPOpenview) and SUN (SunNetmanager), have well-developed software packages for network management, while most vendors support simple telnet or menu-based management interfaces. These protocols are supported over the Internet, enabling a network manager to roam at will, literally around the world, and still have access to a device.

3. Reliable Management Access. Corporate and campus networks have become very highly scrutinized. In most larger networks, 24-hour-a-day maintenance and monitoring takes place to ensure the network is running properly. Networking protocols designed for data delivery ensure that information arrives from node to node. Routed networks provide multiple pathways for data delivery. New software capable of measuring QoS helps the network manager to tune the network topology to allow data to flow freely between devices virtually all the time. All of these reasons combine to make management over the network one of the most reliable ways to manage a remote device.

4. Lower Management Costs. With a reliable remote management tool available, network managers can streamline their staffing and troubleshooting requirements to a centralized or even automated system. Standards-based management features such as SNMP maximize the investment in software and analysis devices based on that protocol. Even a simple management technique such as a ping or a telnet login to validate that a node is alive can be run from a script. With a management scheme based on established standards, network managers can train internal staff better, and can more easily hire new staff with known levels of skill regarding the management suite. Better management technology and better staff result in lower costs for the network manager.

The availability of smaller, more tightly integrated circuits makes it possible to build a single-port device server on a circuit board no larger than a matchbook. As a board-level product, a device server can be integrated into a device's design to allow the following options:

1. For devices that are able to preprocess in full the status and configuration information, one can use the device server to allow that device to be offered with a network port that can accept or generate connections to any network node.

2. For those devices requiring processing elsewhere, a device server can allow that device to be located anywhere on the network, with the serial data being *piped* from the sending device to the processing device.

CHAPTER 6

HOW HOME AND SMALL BUSINESS LANs USE ADDRESSING

Any home PC or workstation that runs the IP protocol by default has its own IP address, just as homes have specific numbers along a street. An IP address is a logical address that is independent of a host's hardware. IP addressing is perhaps the most mystifying part of IP for people who are new to the networking world; however, it really is very simple. To understand, you just need to use a little bit of binary (base 2) arithmetic and a little bit of decimal (base 10) arithmetic.

IP addresses are 32 bits long, and the normal way of writing them is called dotted-decimal notation. To write an address in dotted-decimal notation, we divide the 32 bits of the address into four 8-bit chunks. Each 8-bit chunk is called an octet or a byte. We then convert the octets from binary to decimal and put dots (.) between them. Figure 6.1 shows four forms of the same IP host address.

The first line in Figure 6.1 is normal binary, just a string of 32 1s and 0s; each 1 and 0 is a bit. This line is difficult for people to read, but it is what a computer, like a router, sees. The decimal representation of the binary address has a rather large value; 32 bits can represent decimal numbers between 0 and 4,294,967,295. The second line of Figure 6.1 is the decimal equivalent of the binary line. How would you like to read a number like that every time you wanted to communicate a host address to someone?

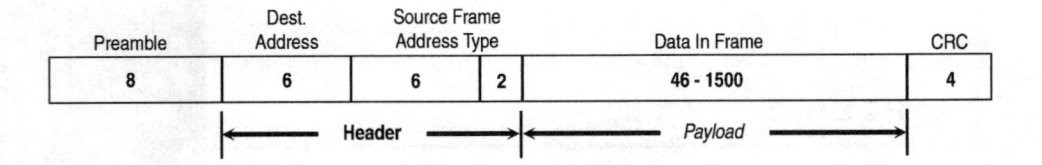

Preamble	Dest. Address	Source Address	Frame Type	Data In Frame	CRC
8	6	6	2	46 - 1500	4

←———— Header ————→←———— Payload ————→

FIGURE 6.1 How IP Host Addressing Works

The third line of Figure 6.1 is just an intermediate step toward the last line, with the 32 bits divided into four octets (quarters). The last line, dotted-decimal notation, is the one that we use, and it is the easiest to read and write.

Because each of the four octets of an IP address is 8 bits and because the decimal values that can be represented with 8 bits range from 0 to 255, the value of any one of the numbers in a dotted-decimal IP address cannot exceed 255.

IP addresses have two main parts—a *network part* that identifies the network where a host resides, and a *node part* that identifies a specific host on the network (sort of like a street name and a house number). The network and node parts together make up the full IP address of a host. The network part is used by routing software to determine for which network a packet is destined. The node part is used by routing software to send a packet to an individual host once the packet has reached the host's network. Just to make this a little more exciting, the line between network and node moves.

Three things can be used to tell which part of an address is network and which part of an address is node:

- Network mask
- Prefix length
- Class

The network mask explicitly specifies which part of an IP address represents a network. The network mask is 32 bits long and is normally written in dotted-decimal notation just like an address. An IP address and mask are usually written together, with the mask immediately after the address.

A network mask has binary 1s in the bit positions that represent the network part of an address, and binary 0s in the bit positions that represent the node part of an address. The binary 1s must start from the left (most-significant) side of the mask and extend contiguously (they must be side by side) until the network part ends; the rest of the mask must be all 0s.

For example, the mask

```
255.255.255.0
```

when paired with an IP address, would tell us that the first 24 bits of the address is considered network and the last 8 bits are node. This becomes clear when the binary equivalent of this mask is examined:

```
11111111  11111111  11111111  00000000  (binary)

255  .  255  .  255  .  0  (dotted-decimal)
```

We see that the mask has 24 1s, starting from the left, and 8 0s. Let's apply this mask to the IP address in Figure 6.1.

```
11000000  10101000  10000001  01100011  (Address)

11111111  11111111  11111111  00000000  (Mask)
```

Because the mask indicates that the first 24 bits of the address are the network, the network address must be this:

```
11000000  10101000  10000001  01100011

192  .  168  .  129  .  0
```

An IP address that has binary 0s in all of the node bits represents a network, not a host. An IP address that has binary 1s in all of the node bits represents all of the hosts on a network; this is called a *broadcast address*. The broadcast address of the 192.168.129.0 network in the above example is this:

```
10101100  00010000  10000001  11111111
```

192 . 168 . 129 . 255

The node values between all 0s and all 1s identify individual hosts on a network. Therefore, the 192.168.129.0 255.255.255.0 network can have hosts with addresses between 192.168.129.1 and 192.168.129.254. There are 254 valid host addresses on this particular network. The equation below shows the simple formula for calculating the number of valid host addresses on a network. Applying this formula to our running example with 8 node bits has the following results:

$2^8 - 2 = 256 - 2 = 254$ Hosts

The answer is the same: 254 hosts on the network 192.168.129.0 255.255.255.0.

Some publications refer to an address's network part—the part that is significant for routing—as a prefix; therefore, all we really have to do is specify the length of the prefix to communicate the same information that the network mask communicates.

SUBNETTING

There is a limited number of IP network addresses, and, as each network must have its own address, we sometimes need to artificially increase the number of networks we can address at the expense of reducing the number of hosts on each network.

Suppose, for example, that we have been assigned a class-B address to address our network. A class-B address has 16 node bits, so we can address a single network that has 65,534 hosts. Our internetwork is likely to have many networks, and none of them is likely to have that many hosts. Therefore, to utilize the assigned address space efficiently, we can take some of the node bits and use them to address networks (subnets) instead of hosts. A subnet is still a network, but it has an address that has been derived from the classification of service it provides on a network.

Subnetting is the act of taking some of the node bits and using them as network bits. This is accomplished by extending the binary 1s of the network mask to the right to include some of the node bits. The length of the mask must still be 32; therefore, to increase the number of network bits, we must decrease the number of node bits.

The extent of the increase in network bits depends on the number of subnets we need and the number of hosts we need on each subnet. Tables 6.1 and 6.2 illustrate this concept. The number of subnets uses the same formula except that subnet bits have been substituted for node bits. The same formulas can be applied to class A addresses, even though they are not shown in the tables. The number of subnet bits in the Tables 6.1 and 6.2 is the number of bits that the mask has been extended to the right from the default according to the address class.

Once we determine what type of subnetting will be used, we can calculate addresses for each of the networks (or subnets). Using the default mask for a class C address, we determine that this is the address of a host on the 192.168.129.0 network. Now suppose that we need to use this same network to address six networks that will have no more than 30 hosts each.

For example, in our mask, the last non-0 octet is the fourth octet, and its value is 224. Subtracting 224 from 256 yields 32. The subnet seed is 32, and all of the subnet addresses are multiples of 32 in the fourth octet. The network addresses of our six subnets are these:

- 192.168.129.32
- 192.168.129.64
- 192.168.129.96
- 192.168.129.128
- 192.168.129.160
- 192.168.129.192

TABLE 6.1 How Network Masks Are Used Define Supported Hosts

Network Mask	Prefix Length	Subnet Bits	Node Bits	Subnets	Hosts
255.255.0.0	/16	0	16	0 (1 Net)	65,534
255.255.192.0	/18	2	14	2	16,382
255.255.224.0	/19	3	13	6	8,190
255.255.240.0	/20	4	12	14	4,094
255.255.248.0	/21	5	11	30	2,046
255.255.252.0	/22	6	10	62	1,022
255.255.254.0	/23	7	9	126	510
255.255.255.0	/24	8	8	254	254
255.255.255.128	/25	9	7	510	126
255.255.255.192	/26	10	6	1,022	62
255.255.255.224	/27	11	5	2,046	30
255.255.255.240	/28	12	4	4,094	14
255.255.255.248	/29	13	3	8,190	6
255.255.255.252	/30	14	2	16,382	2

TABLE 6.2 Node Bits Vary the Number of Supported Hosts for a Subnet Mask

Network Mask	Prefix Length	Subnet Bits	Node Bits	Subnets	Hosts
255.255.255.0	/24	0	8	0 (1 Net)	254
255.255.255.192	/26	2	6	2	62
255.255.255.224	/27	3	5	6	30
255.255.255.240	/28	4	4	14	14
255.255.255.248	/29	5	3	30	6
255.255.255.252	/30	6	2	62	2

The broadcast address of each subnet is the address of the next higher subnet minus 1. For example, the broadcast address of the 192.168.129.96 subnet is 192.168.129.127 (192.168.129.128 minus 1). Remember that the valid host addresses for a network or subnet are those addresses between the network's address and its broadcast address; therefore, the valid host addresses for the 192.168.129.96 255.255.255.224 network range from 192.168.129.97 through 192.168.129.126—that's 30 hosts, just like we calculated.

Our example IP host address—192.168.129.99—is in this range; therefore, it is a host on the 192.168.129.96/127 subnet. We could have simply calculated this by using a calculation similar to the one we used to calculate the subnet addresses. The last non-0 value in the network mask is 224 in the fourth octet. Dividing the fourth octet of the host address (99) by the subnet seed (32) gives a result of 3 (never mind the remainder). Multiplying the result (3) by the seed (32) yields the subnet address's fourth octet value (96).

PUBLIC AND PRIVATE IP ADDRESSES

Network traffic that traverses a public network like the Internet must use public addresses. When we connect one of our networks directly to the Internet, we usually have public addresses assigned to us by an ISP. ISPs get their public addresses from the American Registry for Internet Numbers (ARIN).

IP packets with public addresses can be routed on the Internet because the addresses are unique; however, because of a shortage of public addresses, private addresses have been defined for use on networks that are not going to connect directly to the Internet. RFC 1918 defines one class-A network address, 16 class-B network addresses, and 256 class-C network addresses to be private. These addresses are

- 10.0.0.0
- 172.16.0.0 through 172.31.0.0
- 192.168.0.0 through 192.168.255.0

The IP address that we have been using in our examples—192.168129.99—is taken from a private class-C network address.

Many companies are using these private network addresses to address their internal internetworks (intranetworks) that connect to the Internet through a firewall or some other system that performs Network Address Translation (NAT). In these cases, NAT is used to translate the private addresses found in the internal IP packets into public addresses, to enable the packets to be routed on the external Internet.

IP OVERVIEW

The TCP/IP protocol suite has been around since the late 1970s; this is long before the OSI Reference Model was developed. However, because the OSI Model *is* a *reference model*, we will refer to it when discussing the layers of the IP stack. The OSI Model and the IP stack are shown side by side in Figure 6.2.

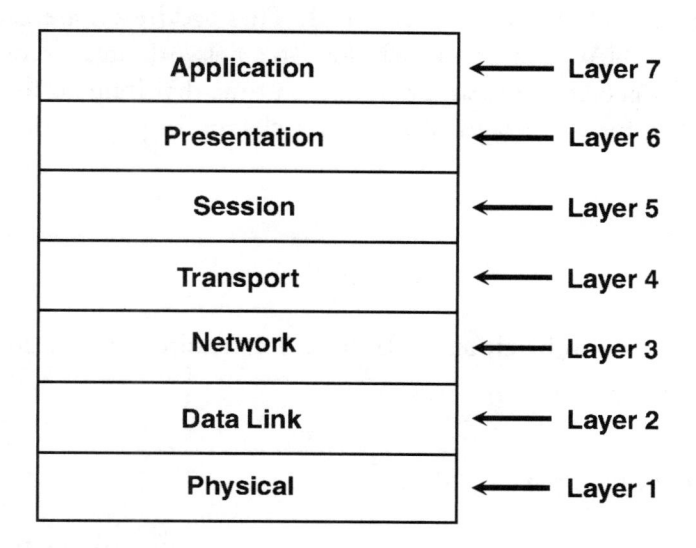

Layer 7	Application
Layer 6	Presentation
Layer 5	Session
Layer 4	Transport
Layer 3	Network
Layer 2	Data Link
Layer 1	Physical

FIGURE 6.2 How the OSI Model Is Constructed

Layers 1 and 2

Layers 1 and 2 deal with the connection of a host to a network. IP running on a host is independent of the type of network connection.

Layer 1 defines the physical medium where a host is attached. A host running IP can be attached to just about any kind of network.

Layer 2 defines the format of the frame header and trailer that a host builds and wraps around a packet before transmitting the packet. The frame header contains a field that indicates the protocol encapsulated between the frame header and the frame trailer. The field is called either protocol type or Service Access Point (SAP) depending on the encapsulation type configured on the host's interface. For example, with ARPA encapsulation on Ethernet, a protocol type field value of 0×0800 (hexadecimal 0800) indicates that an IP packet is encapsulated within the frame and that an IP header immediately follows the frame header. With SAP encapsulation, an SAP field value of 0×06 means the same thing.

The layer 2 header of a frame traveling a LAN contains the addresses that are used to identify the network host that transmitted the frame, and the host, or hosts, to whom the frame is destined. These addresses are called MAC addresses. The MAC address of each host on a network must be unique. The layer 2 header gets removed and rebuilt each time that frame is forwarded by a router; therefore, the layer 2 addresses change in the header from one network to the next.

Layer 3

At layer 3, there can be either an IP header or Address Resolution Protocol (ARP) data.

Internet Protocol

IP is used for the exchange of data between hosts. The IP header is located at layer 3 of an IP packet. The IP header does not contain real data, but it does contain many items that are used by the protocol. Some of these are here:

- Source IP address
- Destination IP address
- Time to live (TTL)
- Protocol number

The source IP address is the 32-bit IP address of the host that originated the packet. The destination IP address is the 32-bit IP address that routers examine to determine where to route a packet. The destination IP address can be a unicast host address, a broadcast address, or a multicast address. Normally, neither the source IP address nor the destination IP address changes from the time a packet leaves a host to the time it arrives at its destination.

The TTL field in a packet's IP header is decremented each time the packet is routed by a router. The use of TTL keeps a packet from being routed endlessly on an internetwork when its destination cannot be found. When a router decrements a packet's TTL to zero (0), the router must drop the packet. The router can then send an ICMP message to the packet's source indicating the reason the packet was dropped (TTL exceeded). The TTL field has a maximum value of 255; therefore, in theory, if the source host sent a packet with a beginning TTL of 255, the packet could get routed 255 times before it got dropped.

The protocol number identifies the next-level protocol—that is, what is after the IP header in the packet. The following are some examples of items that could follow the IP header and their associated protocol numbers:

- TCP header—protocol number 6
- UDP header—protocol number 17
- ICMP message—protocol number 1
- IGRP message—protocol number 9

Upper-layer applications usually use TCP or UDP, but before a host gets to the application data, it must read a TCP or UDP header. TCP and UDP are probably the protocol numbers used the most often.

Address Resolution Protocol

Hosts use ARP to acquire the MAC address of another host to which a frame is being transmitted. This is done during the frame encapsulation process. A source host presumably would already know the IP address of a destination host and would be able to build the layer 3 (IP) header containing both source and destination IP addresses. However, the source host also must build a frame header for encapsulation of the packet on a LAN. The frame header contains the source and destination MAC addresses. The source host knows its own MAC address, but it needs to know the destination host's MAC address to complete the frame header.

The source host locates a destination host's MAC address by sending a layer-2 broadcast frame onto the LAN asking for the MAC address associated with a given IP address; this frame is called an ARP request. Every host on a network examines a layer-2 broadcast frame. If the destination host is on the network, it will recognize the IP address in the ARP request frame and send an ARP reply to the source host. The ARP reply contains the destination host's MAC address that the source host uses to complete the frame header. The source host also puts the destination host's MAC address and IP address into a table, called the *ARP cache*. The next time the source host wants to send a frame to the destination host, it can just look up the destination host's MAC address in the ARP cache—it does not need to send the ARP request.

If a router receives an ARP request for a destination host on a network other than the one on which the request was received, the router can reply on behalf of the destination host. The source host receives an ARP reply containing the router interface's MAC address, thus instructing the source host to send its frame to the router for routing. ARP is extremely important in IP-based networks. Some textbooks call ARP a layer 2 protocol because its messages are never routed at layer 3. ARP is included at layer 3 in this case only because the ARP messages follow the layer 2 header.

Layer 4

A router is not normally concerned with information above layer 3 in a packet unless the router is generating its own traffic, receiving traffic destined for it, or performing packet filtering.

IP traffic generated by a router can include ICMP messages and routing protocol updates. A router both receives and generates traffic when a telnet connection is established to it or from it.

Transmission Control Protocol

TCP is a connection-oriented protocol. This means that applications can use TCP at layer 4 to establish a connection with each other, and the applications can depend on TCP to perform sequencing, acknowledging, and windowing of their data. TCP is considered to be a reliable transport for application data.

Port numbers in the TCP header identify the upper-layer applications, source, and destination that are using the connection.

User Datagram Protocol

UDP is a connectionless protocol. Applications that use UDP at layer 4 must establish a connection themselves if a connection is required. UDP does no data sequencing, acknowledging, or windowing. UDP is used when applications want to handle these tasks themselves or when applications simply want to send data to other applications without the overhead and delay of connection establishment.

Similar to TCP, port numbers in the UDP header identify the upper-layer application from which a message has come and the upper-layer application for which a message is destined.

Internet Control and Messaging Protocol

Hosts (especially routers) use ICMP to communicate network conditions or errors to other hosts. For example, ICMP is used to communicate a packet's exceeded TTL to the host that sent the packet. ICMP messages, specifically ICMP *Destination Unreachable* messages, are also used to communicate broken routing or nonexistent networks to transmitting hosts. Probably the most recognizable use of ICMP is found within the ping application. The IP implementation of ping uses ICMP *echo* messages to test connectivity. When we want to find out if a host is accessible, we can ping the host with an ICMP echo request containing data. If the host receives the request, it should respond by returning an ICMP echo response containing the same data.

Layers 5, 6, and 7

Layers 5, 6, and 7 are loosely bundled for all IP-based applications. Most of the applications we use every day run over TCP or UDP. IOS can run many of these applications.

TCP Applications

Here are some TCP applications that can run on IOS:

- Telnet (client and server)
- rcp (client)
- HTTP (server)

Telnet is a remote session application that is used when configuring a remote router. Rcp is a file copy application that can be used to copy files to and from a Cisco router. HTTP is the application used to transfer World Wide Web (WWW) pages from web servers to web clients (browsers).

UDP Applications

UDP applications that can run on IOS include these:

- TFTP (client)
- SNMP (client)
- DNS (client)

TFTP is a file copy application that is used to copy files to and from a Cisco router. SNMP is a network management application used to monitor and manage network devices. DNS is the application used to translate host names to IP addresses and vice versa.

CHAPTER 7

Exploring the Fundamentals of Transmission Media for Home and Small Businesses

INTRODUCTION

The performance of your home or office network depends on the cabling you choose to use in creating it. Where once there was the vision of home networking having that bright blue cabling running from room to room in a house, today's networking is much more focused on how to make the experience more enjoyable and less obtrusive to your overall lifestyle. Many companies—3Com and DLink among them—see an opportunity to turn millions of homes and offices around the world into small networks. This will give you more options than ever before for getting your own LAN up and running. The best news of all about these developments is that these companies are making convenience a priority in designing their networks.

This chapter discusses some of the most common network transmission media and will help you to define the type(s) of transmission media that will be best for you to use in creating your own LAN, whether it is in your office or at home. One broad classification of this transmission media is known as *bounded media*, or *cable media*. This includes cable types such as coaxial, shielded twisted pair (STP), unshielded twisted pair (UTP), and fiber-optic. Another type of media is known as *boundless media*; these media include all forms of wireless communications. To lay the groundwork for these issues, the chapter begins with an introduction to the frequencies in the electromagnetic spectrum and a look at some important characteristics of the transmission media that utilize these different frequencies to transmit the data.

EXPLORING TYPES OF TRANSMISSION FREQUENCIES

Transmission media make it possible to send electronic signals from one computer to another. These electronic signals express data values in the form of binary (on/off) impulses, which are the basis for all computer information (represented as 1s and 0s). These signals are transmitted between the devices on the network, using some form of transmission media (such as cables or radio), until they reach the desired destination computer.

All signals transmitted between computers consist of some form of electromagnetic (EM) waveform, ranging from radio frequencies to microwaves and infrared light. Different media are used to transmit the signals, depending on the frequency of the EM waveform. The EM spectrum consists of several categories of waveforms, including radio frequency waves, microwave transmissions, and infrared light.

The frequency of a wave is dependent upon the number of waves or oscillations that occur during a period of time. An example that most people can relate to is the difference between a high-pitched sound such as a whistle—very high frequency, with numerous cycles of oscillation (or waves) each second—and a low-pitched sound such as a foghorn. Radio frequency waves are often used for LAN signaling and can be transmitted across electrical cables (twisted pair or coaxial) or by radio broadcast.

Microwave transmissions can be used for tightly focused transmissions between two points. For example, they are used to communicate between earth stations and satellites, as well as for line-of-sight transmissions on the earth's surface. In addition, microwaves can be used in low-power forms to broadcast signals from a transmitter to many receivers. Cellular phone networks are examples of systems that use low-power microwave signals to broadcast signals.

Infrared light is ideal for many types of network communications. IR light can be transmitted across relatively short distances and can either be beamed between two points or be broadcast from one point to many receivers. IR and higher frequencies of light also can be transmitted through fiber-optic cables. A typical television remote control uses infrared transmission.

DISCOVERING TRANSMISSION MEDIA CHARACTERISTICS

Each type of transmission media has special characteristics that make it suitable for a specific type of service. In constructing your own home network, you need to be aware of the following characteristics that can be used as the basis for comparing various transmission media:

- Cost
- Installation requirements
- Bandwidth
- Band usage (baseband or broadband)
- Attenuation
- Immunity from electromagnetic interference

COST

Cost is definitely one of the most important considerations when you are creating a LAN for your home or office. You'll be weighing the benefits of the fastest possible speeds for your network against the costs associated with state-of-the-art cabling. Mainstream, commonly used transmission media is cheaper but slower. As with nearly everything else in the computer field, the fastest technology is the newest, and the newest is the most expensive. Over time, economies of scale will bring the price down, but by then a newer technology will have come along.

INSTALLATION REQUIREMENTS

Installation requirements typically involve two factors. One is that some transmission media require skilled labor to install, and bringing in a skilled outside technician to make changes to or replace resources on the network can bring about undue delays and costs. The second factor has to do with the actual physical layout of the network. Some types of transmission media install more easily over areas where people are spread out, whereas other transmission media are easier to bring to clusters of people or a roaming user.

BANDWIDTH

In computer networking, the term *bandwidth* refers to the measure of the capacity of a medium to transmit data. For example, a medium that has a high capacity has a high bandwidth, whereas a medium that has limited capacity has a low bandwidth

> **Note:** The term *bandwidth* also has another meaning. In the communications industry, bandwidth refers to the range of available frequencies between the lower frequency limit and the upper frequency limit. Frequencies are measured in Hertz (Hz), or cycles per second. The bandwidth of a voice telephone line is 400–4,000 Hz, which means that the line can transmit signals with frequencies ranging from 400 to 4,000 cycles per second.

Bandwidth can be best explained by using water hoses as an analogy. If a 1/2-inch garden hose can carry water flow from a trickle up to 2 gallons per minute, then that hose can be said to have a bandwidth of 2 gallons per minute. A 4-inch fire hose, however, might have a bandwidth that exceeds 100 gallons per minute.

Data transmission rates are frequently stated in terms of the bits that can be transmitted per second. An Ethernet LAN theoretically can transmit 10 million bits per second and has a bandwidth of 10 Mbps.

The bandwidth that a cable can accommodate is determined in part by the cable's length. A short cable generally can accommodate greater bandwidth than a long cable, which is one reason all cable designs specify maximum lengths for cable runs. Beyond those limits, the highest-frequency signals can deteriorate, and errors begin to occur in data signals. You can actually see this by taking a garden hose and snapping it up and down. As the waves travel down the hose, they get smaller as they get farther from your hand. This loss of the wave's amplitude represents attenuation, or signal degradation.

> **Note:** As you know, everything in computers is represented with 1s and 0s. We use 1s and 0s to represent the bits in the computer. However, be sure to remember that transmission media is measured in mega*bits* per second (Mbps), not mega*bytes* per second (MBps). The difference is eightfold—there are 8 bits in a byte.

BAND USAGE (BASEBAND OR BROADBAND)

The two ways to allocate the capacity of transmission media are with baseband and broadband transmissions. Baseband devotes the entire capacity of the medium to one communication channel. Broadband enables two or more communication channels to share the bandwidth of the communications medium. Baseband is the most common mode of operation. For example, most LANs function in baseband mode. Baseband signaling can be accomplished with both analog and digital signals.

Although you might not realize it, you have a great deal of experience with broadband transmissions. Consider, for example, that the TV cable coming into your house from an antenna or a cable provider is a broadband medium. Many television signals can share the bandwidth of the cable because each signal is modulated using a separately assigned frequency. You can use the television tuner to select the frequency of the channel you want to watch.

Multiplexing

Multiplexing is a technique that enables broadband media to support multiple data channels. Multiplexing makes sense under a number of circumstances:

- *When media bandwidth is costly.* A high-speed leased line, such as a T1 or T3, is expensive to lease. If the leased line has sufficient bandwidth, multiplexing can enable the same line to carry mainframe, LAN, voice, video conferencing, and various other data types.

- *When bandwidth is idle.* Many organizations have installed fiber-optic cable that is used to only partial capacity. With the proper equipment, a single fiber can support hundreds of megabits—even a gigabit or more—of data per second.

- *When large amounts of data must be transmitted through low-capacity channels.* Multiplexing techniques can divide the original data stream into several lower-bandwidth channels, each of which can be transmitted through a lower-capacity medium. The signals then can be recombined at the receiving end.

Attenuation

Attenuation is one of the reasons that cable designs must specify limits in the lengths of cable runs. When signal strength falls below certain limits, the electronic equipment receiving the signal can experience difficulty in isolating the original signal from the noise present in all electronic transmissions. The effect is exactly like what happens when you try to tune in distant radio signals—even if you can lock on to the signal on your radio, the sound generally still contains more noise than the sound for a local radio station. As mentioned in chapter 4, repeaters are used to regenerate signals; hence, one solution in dealing with attenuation is to add a repeater.

Electromagnetic Interference

Electromagnetic interference (EMI) consists of outside electromagnetic noise that distorts the signal in a medium. When you listen to an AM radio, for example, you often hear EMI in the form of noise caused by nearby motors or lightning. Some network media are more susceptible to EMI than others.

Cross talk is a special kind of interference caused by adjacent wires, and it occurs when the signal from one wire is picked up by another wire. You may have experienced this when talking on a telephone and hearing another conversation going on in the background. Cross talk is a particularly significant problem with computer networks because large numbers of cables are often located close together, with minimal attention to exact placement.

COMPARING CABLE MEDIA

For the task of creating your home or office LAN you need to know how to make decisions about network transmission media based on some of the factors already described in this chapter. This section discusses three types of network cabling media (wireless technologies will also be discussed later in this chapter):

- Coaxial cable

- Twisted-pair cable

- Fiber-optic cable

Coaxial Cable

Coaxial cables were the first cable types used in LANs. Coaxial (or coax) cable gets its name from the fact that two conductors share a common axis. A type of coaxial cable that you may be familiar with is your television cable.

A coaxial cable has four components, illustrated in Figure 7.1.

- A center conductor, although usually solid copper wire, is sometimes made of stranded wire.

- An outer conductor forms a tube surrounding the center conductor. This conductor can consist of braided wires, metallic foil, or both. The outer conductor, frequently called the shield, serves as a ground and also protects the inner conductor from EMI.

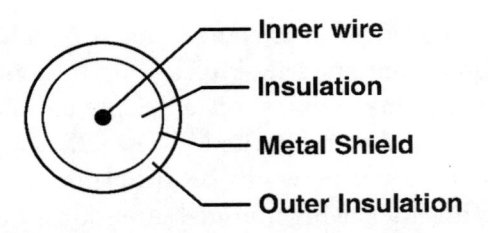

FIGURE 7.1 The Structure of Coaxial Cable

- An insulation layer keeps the outer conductor spaced evenly from the inner conductor.

- A plastic encasement (jacket) protects the cable from damage.

Note: All coaxial cables have a characteristic measurement called *impedance*, which is measured in ohms. Impedance is a measure of the apparent resistance to an alternating current. You must use a cable that has the proper impedance in any given situation.

TYPES OF COAXIAL CABLE

The two basic classifications for coaxial cable are

- Thinnet

- Thicknet

THINNET. Thinnet is a light and flexible cabling medium that is inexpensive and easy to install (see Table 7.1 for Thinnet classifications). Note that Thinnet falls under the RG-58 family, which has a 50-ohm impedance. Thinnet is approximately 6 millimeters (mm)(0.25 inches) in thickness and can reliably transmit a signal for 185 meters (about 610 feet).

THICKNET. Thicknet is thicker than Thinnet (big surprise). Thicknet coaxial cable is approximately 0.5 inches (13 mm) in diameter. Because it is thicker and does not bend as readily as Thinnet, Thicknet cable is harder to work

TABLE 7.1 Thinnet Cable

Classifications Cable	Description	Impedance (ohms)
RG-58/U	Solid copper center	50
RG-58 A/U	Wire strand center	50
RG-58 C/U	Military version of RG-58 A/U	50
RG-59	Cable TV wire	75
RG-62 AR	Cnet specification	93

with. A thicker center core, however, means that Thicknet can carry more signals for a longer distance than Thinnet. Thicknet can transmit a signal approximately 500 meters (1,650 feet).

Thicknet cable, sometimes called standard Ethernet, can be used to connect two or more small Thinnet LANs into a larger network. Because of its greater size, Thicknet is also more expensive than Thinnet. However, Thicknet can be installed relatively safely outside, running from building to building.

COAXIAL CHARACTERISTICS

In this section you'll get an overview of the installation, cost, bandwidth considerations, and EMI resistance of coaxial cabling.

INSTALLATION. Coaxial cable is typically installed in two configurations: daisy chain (from device to device—Ethernet) and star (ARCNet). The daisy chain is shown in Figure 7.2. The Ethernet cabling shown in Figure 7.3 is an example of Thinnet, which uses RG-58 type cable. Devices connect to the cable by means of T-connectors. Cables are used to provide connections between T-connectors. One characteristic of daisy-chain cabling is that a special connector called a *terminator* must terminate the ends of the cable run.

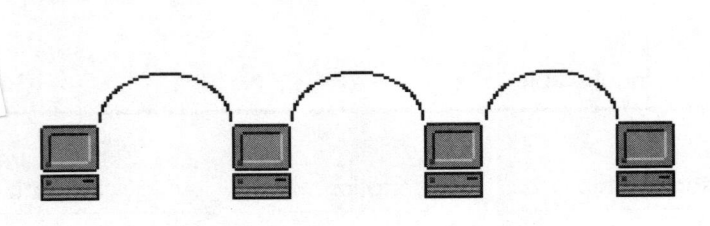

FIGURE 7.2 Daisy-Chain Wiring Configuration

The terminator contains a resistor that is matched to the characteristics of the cable. The resistor prevents signals that reach the end of the cable from bouncing back and causing interference.

Coaxial cable is reasonably easy to install because the cable is robust and difficult to damage. In addition, connectors can be installed with inexpensive tools and a bit of practice. However, the device-to-device or daisy-chain cabling approach can be difficult to reconfigure when new devices must be installed but cannot be installed near an existing cabling path.

Cost. The coaxial cable used for Thinnet falls at the low end of the cost spectrum, whereas Thicknet is among the more costly options.

Capacity. LANs that employ coaxial cable typically have a bandwidth between 2.5 Mbps (ARCNet) and 10 Mbps (Ethernet). Thicker coaxial cables offer higher bandwidth, and the potential bandwidth of coaxial is about 1000 Mbps. Current LAN technologies, however, don't take advantage of this potential.

EMI Characteristics. All copper media are sensitive to EMI, although the shield in coax makes the cable fairly resistant. Coaxial cables, however, do radiate a portion of their signal, and electronic eavesdropping equipment can detect this radiated signal.

Connectors for Coaxial Cable

Two types of connectors are commonly used with coaxial cable, depending on whether the cable is Thinnet or Thicknet.

The most common connector is the British Naval Connector (BNC) (see Figure 7.3 for examples).

Here are some of the key issues involving connectors used with Thinnet cabling:

- A BNC T-connector connects the network board in the PC to the network. The T-connector attaches directly to the network board.

- BNC cable connectors attach cable segments to the T-connectors.

- A BNC barrel connector connects to Thinnet cables.

- Both ends of the cable must be terminated. A BNC terminator is a special connector that includes a resistor that is carefully matched to the characteristics of the cable system.

- One of the terminators must be grounded. A wire from the connector is attached to a grounded point, such as the center screw of a grounded electrical outlet.

In contrast, Thicknet uses N-connectors, which screw on rather than using a twist lock (see Figure 7.4). As with Thinnet, both ends of the cable must be terminated, and one end must be grounded.

The majority of PCs don't connect directly to the Thicknet cable. Instead, a connecting device called a transceiver is attached to the Thicknet cable. This transceiver has a port for an Attachment Unit Interface (AUI) connec-

FIGURE 7.3 Connectors Used with Thinnet Cable

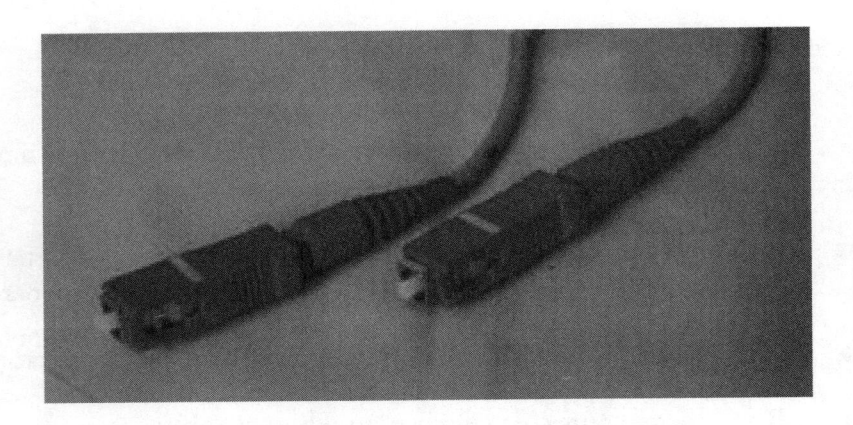

FIGURE 7.4 Connectors Used with Thicknet

tor (which looks deceivingly like a joystick connector), and an AUI cable (also called a *transceiver cable* or a *drop cable*) connects the workstation to the Thicknet medium. Transceivers can connect to Thicknet cables in two ways:

Cutting the cable and splicing N-connectors and a T-connector on the transceiver can connect transceivers; however, because it is so labor intensive, this original method of connecting is used rather infrequently. The more common approach is to use a clamp-on transceiver, which has pins that penetrate the cable without the need for cutting it. Because clamp-on transceivers force sharp teeth into the cable, they frequently are referred to as *vampire taps*.

Note: AUI port connectors are sometimes called DB-15 connectors.

COAX AND FIRE CODE CLASSIFICATIONS

The space above a drop ceiling (between the ceiling and the floor of a building's next level) is extremely significant to both network administrators and fire marshals. This space is a convenient place to run network cables around a building. The plenum (as this space is called), however, is typically an open space in which air circulates freely; consequently, fire marshals pay special attention to it.

The most common outer covering for coaxial cabling is polyvinyl chloride (PVC). PVC cabling gives off poisonous fumes when it burns. For that reason, fire codes prohibit PVC cabling in the plenum because poisonous fumes in the plenum can circulate freely throughout the building.

Plenum-grade coaxial cabling is specially designed to be used without conduit in plenums, walls, and other areas where fire codes prohibit PVC cabling. Plenum-grade cabling is less flexible and more expensive than PVC cabling, so it is used primarily where PVC cabling can't be used.

Twisted-Pair Cable

Twisted-pair cable has become the dominant cable type for all new network designs that employ copper cable. Among the several reasons for the popularity of twisted-pair cable, the most significant is its low cost. Twisted-pair cable is inexpensive to install and offers the lowest cost per foot of any cable type. Your telephone cable is an example of a twisted-pair cable.

A basic twisted-pair cable consists of two strands of copper wire twisted together (see Figure 7.5). The twisting reduces the sensitivity of the cable to EMI. Twisting also reduces the tendency of the cable to radiate radio frequency noise that interferes with nearby cables and electronic components,

FIGURE 7.5 Twisted-Pair Cabling

because the radiated signals from the twisted wires tend to cancel each other out. (Antennas, which are purposely designed to radiate radio frequency signals, consist of parallel, not twisted, wires.)

Twisting of the wires also controls the tendency of the wires in the pair to cause EMI in each other. As noted previously, whenever two wires are in close proximity, the signal in one wire tends to produce cross talk in the other and vice versa. Twisting the wires in the pair reduces cross talk in much the same way that twisting reduces the tendency of the wires to radiate EMI.

SHIELDED TWISTED-PAIR CABLE. STP cabling consists of one or more twisted pairs of cables enclosed in a foil wrap and woven copper shielding. Figure 7.6 shows IBM Type-1 cabling, the first cable type used with IBM Token Ring. Early LAN designers used shielded twisted-pair cable because the shield performed double duty, reducing the tendency of the cable to radiate EMI and reducing the cable's sensitivity to outside interference.

Coaxial and STP cables use shields for the same purpose. The shield is connected to the ground portion of the electronic device to which the cable is connected. A ground is a portion of the device that serves as an electrical reference point and, usually, it is literally connected to a metal stake driven into the ground. A properly grounded shield prevents signals from getting into or out of the cable.

FIGURE 7.6 Shielded Twisted-Pair Cable

The STP example shown in Figure 7.6 (IBM Type 1) includes two twisted pairs of wire within a single shield. Various types of STP cable exist, some that shield each pair individually and others that shield several pairs. The engineers who design a network's cabling system choose the exact configuration. IBM designates several STP cable types to use with their Token Ring network design, and each cable type is appropriate for a given kind of installation. A completely different type of STP is the standard cable for Apple's AppleTalk network.

Because there are so many different types of STP cable, describing precise characteristics is difficult. However, here are some general guidelines:

- **Cost.** STP cable costs more than thin coaxial or UTP cable, but it is less costly than thick coax or fiber-optic cable.

- **Installation.** Naturally, different network types have different installation requirements. One major difference is the connector used. Apple LocalTalk connectors generally must be soldered during installation, a process that requires some practice and skill on the part of the installer.

    In many cases, installation can be greatly simplified with prewired cables—cables precut to length and installed with the appropriate connectors. You must learn to install the required connectors, however, when your installation requires the use of bulk cable. The installation of cables has been regulated or made part of building codes in some areas, to be performed only by a certified cable installer. You should check the regulations regarding this in your area before beginning the installation of any cable.

Unshielded Twisted-Pair Cable

Although UTP cable doesn't incorporate a braided shield into its structure, its characteristics are similar in many ways to STP, differing primarily in attenuation and EMI. UTP pairs are typically color coded to distinguish them.
Telephone systems commonly use UTP cabling. Network engineers can sometimes use existing UTP telephone cabling for network cabling if it is new enough and of a high-enough quality to support network communications.

UTP cable is a latecomer to high-performance LANs because engineers only recently solved the problems of managing radiated noise and susceptibility to EMI. Now, however, there is a clear trend toward using UTP cable, and all new copper-based cabling schemes are based on UTP.

UTP cable is available in the following five grades, or categories. The price of the cable increases as you move from Category 1 to Category 5.

- **Categories 1 and 2.** These voice-grade cables are suitable only for voice and for low data rates (below 4 Mbps). Category 1 was once the standard voice-grade cable for telephone systems. Because of the growing need for data-ready cabling systems, cable in Categories 1 and 2 have been supplanted by Category 3 for new installations.

- **Category 3.** As the lowest data-grade cable, this type of cable is generally suited for data rates up to 10 Mbps, although some innovative schemes utilizing new standards and technologies enable Category 3 cable to support data rates up to 100 Mbps. Category 3, which uses four twisted pairs with three twists per foot, is now the standard cable used for most telephone installations.

- **Category 4.** This data-grade cable, which consists of four twisted pairs, is suitable for data rates up to 16 Mbps.

- **Category 5.** This data-grade cable, also consisting of four twisted pairs, is suitable for data rates up to 100 Mbps. Most new cabling systems for 100-Mbps data rates are designed around Category 5 cable.

In a UTP cabling system, cable is only one component of the system. All connecting devices are also graded, and the overall cabling system supports only the data rates permitted by the lowest-grade component in the system. In other words, if you require a Category 5 cabling system, all connectors and connecting devices must be designed for Category 5 operation.

In addition, there are more stringent requirements regarding installation procedures for Category 5 cable than the lower cable categories. Installers of Category 5 cable require special training and skills to understand these more rigorous requirements.

UTP cable offers an excellent balance between cost and performance characteristics, as discussed in the following sections.

COST. UTP cable is the least costly of any cable type, although properly installed Category 5 tends to be fairly expensive. In some cases, existing cable in buildings can be used for LANs, although you should verify the category of the cable and know the length of the cable in the walls. Distance limits for voice cabling are much less stringent than for data-grade cabling.

INSTALLATION. UTP cable is easy to install. Some specialized equipment might be required, but the equipment is low in cost and its use can be mastered with a bit of practice. Properly designed UTP cabling systems can be reconfigured easily to meet changing requirements. As noted earlier, however, Category-5 cable has stricter installation requirements than lower categories of UTP. Special training is recommended for dealing with Category 5 UTP.

CAPACITY. The data rates possible with UTP have risen from 1 Mbps, past 4 and 16 Mbps, to the point where 100-Mbps data rates are now common.

ATTENUATION. UTP cable shares similar attenuation characteristics with other copper cables. UTP cable runs are limited to a few hundred meters, with 100 meters (a little more than 300 feet) as the most frequent limit.

EMI CHARACTERISTICS. Because UTP cable lacks a shield, it is more sensitive to EMI than coaxial or STP cables. The latest technologies make it possible to use UTP in the vast majority of situations, provided that reasonable care is taken to avoid electrically noisy devices such as motors and fluorescent lights. Nevertheless, UTP might not be suitable for noisy environments such as factories. Cross talk between nearby unshielded pairs limits the maximum length of cable runs.

Fiber-Optic Cable

In almost every way, fiber-optic cable is the ideal cable for data transmission. Not only does this type of cable accommodate extremely high bandwidths, but it also presents no problems with EMI, and it supports durable cables and cable runs as long as several kilometers. The two disadvantages of fiber-optic cable are cost and installation difficulty. Despite these disadvantages, telephone companies now often install fiber-optic as the cable of choice cable into buildings.

The center conductor of a fiber-optic cable is a fiber that consists of highly refined glass or plastic designed to transmit light signals with little loss. A glass core supports a longer cabling distance, but a plastic core is typically easier to work with. The fiber is coated with a cladding or a gel that reflects signals back into the fiber to reduce signal loss. A plastic sheath protects the fiber (see Figure 7.7).

A fiber-optic network cable consists of two strands separately enclosed in plastic sheaths—one strand sends and the other receives. Two types of cable configurations are available: loose and tight. Loose configurations incorporate a space between the fiber sheath and the outer plastic encasement; this space is filled with a gel or other material. Tight configurations contain strength

FIGURE 7.7 Fiber-Optic Cable

wires between the conductor and the outer plastic encasement. In both cases, the plastic encasement must supply the strength of the cable, while the gel layer or strength wires protect the delicate fiber from mechanical damage. Fiber-optic cables don't transmit electrical signals. Instead, the data signals must be converted into light signals. Light sources include lasers and light-emitting diodes (LEDs). LEDs are inexpensive but produce a fairly poor quality of light that is suitable for only less-stringent applications. The end of the cable that receives the light signal must convert the signal back into an electrical form. Several types of solid-state components can perform this service. One of the significant difficulties confronted when installing fiber-optic cable arises when two cables must be joined. The small cores of the two cables (some are as small as 8.3 microns) must be lined up with extreme precision to prevent excessive signal loss.

Fiber-Optic Characteristics

As with all cable types, fiber-optic cables have their share of advantages and disadvantages.

Cost. The cost of fiber-optic cable and the connectors used with it has fallen significantly in recent years. However, the electronic devices required are significantly more expensive than comparable devices for copper cable. Fiber-optic cable is also the most expensive cable type to install.

Installation. Greater skill is required to install fiber-optic cable than to install most copper cables. Improved tools and techniques have reduced the training required, but fiber-optic cable still requires greater care because the cables must be treated fairly gently during installation. For example, every cable has a minimum bend radius, and fibers get damaged if the cables are bent too sharply. It is also important to not stretch the cable during installation.

Capacity. Fiber-optic cable can support high data rates (as high as 200,000 Mbps) even with long cable runs. While UTP cable runs are limited to less than 100 meters with 100-Mbps data rates, fiber-optic cables can transmit 100-Mbps signals for several kilometers.

ATTENUATION. Attenuation is much lower in fiber-optic cables than it is in copper cables. Fiber-optic cables are capable of carrying signals for several kilometers.

EMI CHARACTERISTICS. Because fiber-optic cables don't use electrical signals to transmit data, they are totally immune to EMI. The cables are also immune to a variety of electrical effects that must be taken into account when designing copper cabling systems.

Because the signals in fiber-optic cable are not electrical in nature, the electronic eavesdropping equipment that detects EM radiation cannot detect them. Therefore, fiber-optic cable is the perfect choice for high-security networks.

LEARNING ABOUT WIRELESS MEDIA

The extraordinary convenience of wireless communications has placed an increased emphasis on wireless networks in recent years. Technology is expanding rapidly and will continue to expand into the near future, offering more and better options for wireless networks.

> **Note:** Wireless *point-to-point communications* are another facet of wireless LAN technology. Rather than attempting to achieve an integrated networking capability, wireless point-to-point technology specifically facilitates communications between a pair of devices. For instance, a point-to-point connection might transfer data between a laptop and a home-based computer or between a computer and a printer. Point-to-point signals, if powerful enough, can pass through walls, ceilings, and other obstructions. A point-to-point connection provides data transfer rates of 1.2 to 38.4 Kbps for a range of up to 60.96 meters (200 feet) indoors (or 535 meters [about one-third of a mile] for line-of-sight broadcasts).

Presently, you can subdivide wireless networking technology into three basic types corresponding to three basic networking scenarios:

- **LANs.** Occasionally you will see a fully wireless LAN, but more typically one or more wireless machines function as members of a cable-based LAN.

- **Extended local networks.** A wireless connection serves as a backbone between two LANs. For instance, a company with office networks in two nearby but separate buildings could connect those networks using a wireless bridge.

- **Mobile computing.** A mobile machine connects to the home network using cellular or satellite technology.

REASONS FOR USING WIRELESS NETWORKS

Wireless networks are especially useful for the following situations:

- **Spaces where cabling would be impossible or inconvenient.** These include open lobbies, inaccessible parts of buildings, older buildings, historical buildings where renovation is prohibited, and outdoor installations.

- **Work environments in which people move around a lot.** Network administrators, for instance, must troubleshoot a large office network. Nurses and doctors need to make rounds at a hospital.

- **Temporary installations.** This includes situations like a temporary department that is set up for a specific purpose and then torn down or relocated later.

- **Highly mobile workforces.** These workforces have people who travel outside of the work environment and need instantaneous access to network resources.

- **Geographically remote locations.** These might include satellite offices or branches, ships in the ocean, or teams in remote field locations that need to be connected to a main office or location.

Wireless Communications with LANs

For some of the reasons described earlier in this chapter, it is often advantageous for a network to include some wireless nodes. Typically, though, the wireless nodes are part of what is otherwise a traditional, cable-based network.

An *access point* is a stationary transceiver connected to the cable-based LAN that enables the cordless PC to communicate with the network. The access point acts as a conduit for the wireless PC. The process is initiated when the wireless PC sends a signal to the access point; from there, the signal reaches the network. Therefore, the truly wireless communication is the communication from the wireless PC to the access point. This is similar to when you use your remote control for your TV. Think of the remote control unit in your hand as the computer and the area on the TV set that receives the signal as your access point, or stationary receiver. Use of an access point transceiver is one of several ways to achieve wireless networking. Some of the others are described in later sections.

You can classify wireless LAN communications according to transmission method. The most common LAN wireless transmission methods are these:

- Infrared
- Laser
- Narrowband radio
- Spread-spectrum radio
- Microwave

The following sections look briefly at these important wireless transmission methods. Because of vast differences in evaluation criteria such as costs, ease of installation, range, and EMI characteristics, these items are evaluated in Table 7.2. (Bandwidth usage is not evaluated because wireless media is not a bound communication medium.)

TABLE 7.2 U.S. Home Networking Technology Specifications

Technology	Ethernet	Phone Line	Power Line	HomeRF	802.11	Bluetooth
Current data rate (Mbps)	10/100	10	350 Kbps	2	11	0.72
Future data rate (Mbps)	NA	100	10	10	NA	10 or 20
Availability	Now	Now	Now	Now	Now	2H00
Proponent		HomePNA	HomePlug Powerline Alliance			
Pros	Secure, reliable, and fast	Many backers	Easy to use, flexible access	Mobile, easy to use	Fast data transmissions	Will be used in multiple devices
	One industry standard	Endorsed by top OEMs	Available; power lines ubiquitous	Voice applications	Mobile, easy to use	Lower cost
		Easy integration into other silicon	Lower cost	Lower cost		
Cons	Requires new wiring	Phone jacks not ubiquitous	Multiple standards, competing technologies	Limited range	More expensive	Currently not robust enough to be a full home network
			HomePlug must choose standard quickly	Slower data throughput		Product not yet available
			Slow speeds, subject to interference			

Source: e-Market Dynamics, 2000

INFRARED TRANSMISSION

You use an IR communication system every time you control your television with a remote control. When you push the buttons on it, the remote control transmits pulses of IR light that carry coded instructions to a receiver on the TV. This technology also is used for network communication.

There are four varieties of IR communications:

- **Broadband optical telepoint.** This method uses broadband technology. Data transfer rates in this high-end option are competitive with those for a cable-based network.

- **Line-of-sight IR.** Transmissions must occur over a clear line-of-sight path between transmitter and receiver.

- **Reflective IR.** Wireless PCs transmit toward a common central unit, which then directs communications to each of the nodes.

- **Scatter IR.** Transmissions reflect off of floors, walls, and ceilings until (theoretically) they finally reach the receiver. Because of the imprecise trajectory, data transfer rates are slow. The maximum reliable distance is around 100 feet.

IR transmissions are typically limited to 100 feet or less but, within this range, IR is relatively fast. Its high bandwidth supports transmission speeds of up to 10 Mbps. IR devices are insensitive to RF interference, but reception can be degraded by bright light. Because transmissions are tightly focused, they are fairly immune to electronic eavesdropping. IR transmissions are commonly used for LAN transmissions, yet can also be employed for Wide Area Network (WAN) transmissions as well.

LASER TRANSMISSION

High-powered laser transmitters can transmit data for several thousand yards when line-of-sight communication is possible. Lasers can be used in many of the same situations as microwave links (described later in this chapter), but they do not require an FCC license. On a LAN scale, laser light technology is similar to IR technology. Laser light technology is employed in both LAN and WAN transmissions, although it is more commonly used in WAN transmissions.

Narrowband Radio Transmission

In narrowband radio communications (also called single-frequency radio), transmissions occur at a single RF. The range of narrowband radio is greater than that of IR, effectively enabling mobile computing over a limited area. Neither the receiver nor the transmitter must be placed along a direct line of sight—the signal can bounce off of walls, buildings, and even the atmosphere; but heavy walls, such as steel or concrete enclosures, can block the signal.

Spread-Spectrum Radio Transmission

Spread-spectrum radio transmission is a technique originally developed by the military to solve several communication problems. Spread-spectrum radio improves reliability, reduces sensitivity to interference and jamming, and is less vulnerable to eavesdropping than single-frequency radio. Spread-spectrum radio transmissions are commonly used for WAN transmissions that connect multiple LANs or network segments together. As its name suggests, spread-spectrum transmission uses multiple frequencies to transmit messages. Two techniques employed are frequency hopping and direct-sequence modulation. Spread-spectrum radio transmissions are often used to connect multiple LAN segments together; thus it is often a WAN connection.

> **Note:** Wireless technology can connect LANs in two different buildings into an extended LAN. Of course, this capability is also available through other technologies but, depending on the conditions, a wireless solution is sometimes more cost effective. A wireless connection between two buildings (called a *wireless bridge*) also provides a solution to the potential ground problem described earlier in this chapter.

A wireless bridge acts as a network bridge, merging two local LANs over a wireless connection. Wireless bridges typically use spread-spectrum radio technology to transmit data for up to 3 miles. A device called a *long-range wireless bridge* has a range of up to 25 miles.

Microwave Transmission

Microwave technology has applications in all three of the wireless networking scenarios: LAN, extended LAN, and mobile networking. Mobile computing is a growing technology that provides almost unlimited range for traveling computers by using satellite and cellular phone networks to relay the signal to a home network. Mobile computing typically is used with portable PCs or PDA devices.

Three forms of mobile computing are as follows:

- **Packet-radio networking.** The mobile device sends and receives network-style packets via satellite. Packets contain a source and destination address, and only the destination device can receive and read the packet.

- **Cellular networking.** The mobile device sends and receives cellular digital packet data (CDPD) using cellular phone technology and the cellular phone network. Cellular networking provides very fast communications.

- **Satellite station networking.** Satellite mobile networking stations use satellite microwave technology.

SATELLITE MICROWAVE. Satellite microwave systems relay transmissions through communication satellites that operate in geosynchronous orbits 22,300 miles above the earth. Satellites orbiting at this distance remain located above a fixed point on earth.

Earth stations use satellite dishes to communicate with satellites. These satellites then can retransmit signals in broad or narrow beams, depending on the locations set to receive the signals. When the destination is on the opposite side of the earth, for example, the first satellite cannot transmit directly to the receiver and thus must relay the signal through another satellite.

Because no cables are required, satellite microwave communication is possible with most remote sites and with mobile devices, which enables communication with ships at sea and motor vehicles.

The distances involved in satellite communication make for an interesting phenomenon: Because all signals must travel 22,300 miles to the satellite and 22,300 miles when returning to a receiver, the time required to transmit a signal is independent of distance on the ground. It takes the same amount of time to transmit a signal to a receiver in the same state as it does to a receiver a third of the way around the world. The time required for a signal to arrive at its destination is called *propagation delay*. The delays encountered with satellite transmissions range from 0.5 to 5 seconds.

Unfortunately, satellite communication is extremely expensive. Building and launching a satellite can easily cost in excess of a $1 billion. In most cases, organizations share these costs or purchase services from a commercial provider. AT&T, Hughes Network Services, and Scientific Atlanta are among the firms that sell satellite-based communication services.

Satellite links operate in the low GHz range, typically at 11–14 GHz. Costs are extremely high and are usually distributed across many users when communication services are sold. Bandwidth is related to cost, and firms can purchase almost any required bandwidth. Typical data rates are 1–10 Mbps. Properly designed systems also take attenuation into account. Attenuation characteristics depend on frequency, power, and atmospheric conditions (rain and atmospheric conditions might attenuate higher frequencies). Microwave signals also are sensitive to EMI and electronic eavesdropping, so signals transmitted through satellite microwave frequently are encrypted as well.

SUMMARY

Cable media are often cheaper than wireless media, but cable media are also limited in the distances they can cover. Wireless media are often more susceptible to EMI than fiber-optic cable is, but wireless media are not subject to the accessibility and other installation problems faced by cable. In conclusion, each transmission medium should be evaluated in terms of the obstacles that will be faced when trying to relay a signal from one device on the network to another.

CHAPTER 8

ADMINISTRATOR'S GUIDE TO MANAGING A SMALL BUSINESS LAN

Network administration tasks fall into two very different categories: configuration and troubleshooting. Configuration tasks prepare for the expected; they require detailed knowledge of command syntax, but are usually simple and predictable. Once a system is properly configured, there is rarely any reason to change it. The configuration process is repeated each time a new operating system release is installed, but with very few changes.

In contrast, network troubleshooting deals with the unexpected. Troubleshooting frequently requires knowledge that is conceptual rather than detailed. Network problems are usually unique and sometimes difficult to resolve. Troubleshooting is an important part of maintaining a stable, reliable network service.

In this chapter, we discuss the tools you will use to ensure that the network is in good running condition. However, good tools are not enough. No troubleshooting tool is effective if applied haphazardly. Effective troubleshooting requires a methodical approach to the problem, as well as a basic understanding of how the network works. We'll start our discussion by looking at ways to approach a network problem.

APPROACHING A PROBLEM

To approach a problem properly, you need a basic understanding of TCP/IP. The first few chapters of this book discuss the basics of TCP/IP and provide enough background information to troubleshoot most network problems. To understand a network problem, it is important to have knowledge of how TCP/IP routes data through the network, between individual hosts, and between the layers in the protocol stack. However, it usually isn't necessary to have detailed knowledge of each protocol. When you need these details, look them up in a definitive reference—don't try to recall them from memory.

Not all TCP/IP problems are alike, and not all problems can be approached in the same manner. But the key to solving any problem is understanding what the problem is. This is not as easy as it may seem. The "surface" problem is sometimes misleading, and the "real" problem is frequently obscured by many layers of software. Once you understand the true nature of the problem, the solution to the problem is often obvious.

First, gather detailed information about exactly what's happening. When a user reports a problem, talk to her. Find out which application failed. What is the remote host's name and IP address? What is the user's hostname and address? What error message was displayed? If possible, verify the problem by having the user run the application while you talk her through it. If possible, duplicate the problem on your own system.

Testing from the user's system, as well as from other systems, find out the following:

- Does the problem occur in other applications on the user's host, or is only one application having trouble? If only one application is involved, the application may be misconfigured or disabled on the remote host. Because of security concerns, many systems disable some services.

- Does the problem occur with only one remote host, all remote hosts, or only certain groups of remote hosts? If only one remote host is involved, the problem could easily be with that host. If all remote hosts are involved, the problem is probably with the user's system (particularly if no other hosts on your local network are experiencing the same problem). If only hosts on certain subnets or external networks are involved, the problem may be related to routing.

- Does the problem occur on other local systems? Make sure you check other systems on the same subnet. If the problem occurs only on the user's host, concentrate testing on that system. If the problem affects every system on a subnet, concentrate on the router for that subnet.

Once you know the symptoms of the problem, visualize each protocol and device that handles the data. Visualizing the problem will help you to avoid oversimplification and keep you from assuming that you know the cause even before you start testing. Using your TCP/IP knowledge, narrow your attack to the most likely causes of the problem, but keep an open mind.

TROUBLESHOOTING HINTS

In this section we offer several useful troubleshooting hints. They are not part of a troubleshooting methodology—just good ideas to keep in mind.

- Approach problems methodically. Allow the information gathered from each test to guide your testing. Don't jump on a hunch into another test scenario without ensuring that you can pick up your original scenario where you left off.

- Work carefully through the problem, dividing it into manageable pieces. Test each piece before moving on to the next. For example, when testing a network connection, test each part of the network until you find the problem.

- Keep good records of the tests you have completed and their results. Keep a historical record of the problem in case it reappears.

- Keep an open mind. Don't assume too much about the cause of the problem. Some people believe their network is always at fault, while others assume the remote end is always the problem. Some are so sure they know the cause of a problem that they ignore the evidence of the tests. Don't fall into these traps. Test each possibility and base your actions on the evidence gained from the tests.

- Be aware of security barriers. Security firewalls sometimes block ping, traceroute, and even ICMP error messages. If problems seem to cluster around a specific remote site, find out if they have a firewall.

- Pay attention to error messages. Error messages are often vague, but they frequently contain important hints for solving the problem.

- Duplicate the reported problem yourself. Don't rely too heavily on the user's problem report. The user has probably seen this problem only from the application level. If necessary, obtain the user's data files to duplicate the problem. Even if you cannot duplicate the problem, log the details of the reported problem for your records.

- Most problems are caused by human error. You can prevent some of these errors by providing information and training on network configuration and usage.

- Keep your users informed. This reduces the number of duplicated trouble reports; it also avoids duplication of effort, as in situations where several system administrators might work on the same problem without knowing others are already working on it. If you're lucky, someone may have seen the problem before and have a helpful suggestion about how to resolve it.

- Don't speculate about the cause of the problem while talking to the user. Save your speculations for discussions with your networking colleagues. Your speculations may be accepted by the user as gospel and become rumors. These rumors can cause users to avoid using legitimate network services and may undermine confidence in your network. Users want solutions to their problems; they're not interested in speculative technobabble.

- Stick to a few simple troubleshooting tools. For most TCP/IP software problems, the tools discussed in this chapter are sufficient. Just learning how to use a new tool is often more time consuming than solving the problem with an old familiar tool.

- Thoroughly test the problem at your end of the network before locating the owner of the remote system to coordinate testing with him. The greatest difficulty of network troubleshooting is that you do not always control the systems at both ends of the network. In many cases, you may not even know who *does* control the remote system.

- Don't neglect the obvious. A loose or damaged cable is always a possible problem. Check plugs, connectors, cables, and switches. Small things can cause big problems.

Diagnostic Tools

Because most problems have a simple cause, developing a clear idea of the problem often provides the solution. Unfortunately, this is not always true, so in this section we begin to discuss the tools that can help you attack the most intractable problems. Many diagnostic tools are available, ranging from commercial systems with specialized hardware and software that may cost thousands of dollars to free software that is available from the Internet. Many software tools are provided with your UNIX system. You should also keep some hardware tools handy.

To maintain the network's equipment and wiring, you need some simple hand tools. A pair of needle-nose pliers and a few screwdrivers may be sufficient, but you may also need specialized tools. For example, attaching RJ-45 connectors to UTP cable requires special crimping tools. It is usually easiest to buy a ready-made network maintenance tool kit from your cable vendor.

A full-featured cable tester is also useful. Modern cable testers test both Thinnet and UTP cable, and are small handheld units with a keypad and LCD display. You select the tests from the keyboard and results are displayed on the LCD screen. It is not necessary to interpret the results because the unit does that for you and displays the error condition in a simple text message. For example, a cable test might produce the message, "Short at 74 feet." This tells you that the cable is shorted 74 feet away from the tester. What could be simpler? The proper test tools make it easier to locate, and therefore fix, cable problems.

A laptop computer can be a most useful piece of test equipment when properly configured. Install TCP/IP software on the laptop. Take it to the location where the user reports a network problem. Disconnect the Ethernet cable from the back of the user's system and attach it to the laptop. Configure the laptop with an appropriate address for the user's subnet and reboot it. Then ping various systems on the network and attach to one of the user's servers. If everything works, the fault is probably in the user's computer. The user trusts this test because it demonstrates something she does every day. She will have more confidence in the laptop than she would have in an unidentifiable piece of test equipment displaying the message "No faults found." If the test fails, the fault is probably in the network equipment or wiring. That's the time to bring out the cable tester.

Testing Basic Connectivity

The ping command tests whether a remote host can be reached from your computer. This simple function is extremely useful for testing the network connection, independent of the application in which the original problem was detected. Ping allows you to determine whether further testing should be directed toward the network connection (the lower layers) or the application (the upper layers). If ping shows that packets can travel to the remote system and back, the user's problem is probably in the upper layers. If packets can't make the round-trip, the lower protocol layers are probably at fault.

Frequently a user reports a network problem by stating that he can't telnet (or FTP, or send email, or whatever) to some remote host. He then immediately qualifies this statement with the announcement that it worked before. In cases like this, where the ability to connect to the remote host is in question, ping is a very useful tool.

Using the hostname provided by the user, ping the remote host. If your ping is successful, have the user ping the host. If the user's ping is also successful, concentrate your further analysis on the specific application in which the user is having the problem. Perhaps the user is attempting to telnet to a host that provides only anonymous FTP. Perhaps the host was down when the user tried his application. Have the user try it again while you watch or listen to every detail of what he is doing. If he is doing everything right and the application still fails, you may need to do a detailed analysis of the application with snoop and coordinate with the remote system administrator.

If your ping is successful and the user's ping fails, concentrate testing on the user's system configuration, as well as on those things that are different about the user's path to the remote host in comparison to your path to the remote host.

If your ping fails, or the user's ping fails, pay close attention to any error messages. The error messages displayed by ping are helpful guides for planning further testing. The details of the messages may vary from implementation to implementation, but there are only a few basic types of errors, which we will discuss in the following sections.

UNKNOWN HOST. The remote host's name cannot be resolved by name service into an IP address. The name servers could be at fault (either your local server or the remote system's server), the name could be incorrect, or something could be wrong with the network between your system and the remote server. If you know the remote host's IP address, try to ping that. If you can reach the host using its IP address, the problem is with name service. Use nslookup or dig to test the local and remote servers, as well as to check the accuracy of the host name the user gave you.

NETWORK UNREACHABLE. The local system does not have a route to the remote system. If the numeric IP address was used on the ping command line, reenter the ping command using the hostname. This eliminates the possibility that the IP address was entered incorrectly or that you were given the wrong address. If a routing protocol is being used, make sure it is running, and check the routing table with netstat. If Routing Information Protocol (RIP) is being used, ripquery will check the contents of the RIP updates being received. If a static default route is being used, reinstall it. If everything seems fine on the host, check its default gateway for routing problems.

No Answer. The remote system did not respond. Most network utilities have some version of this message. Some ping implementations print the message "100% packet loss." Telneprints the message "Connection timed out," and sendmail returns the error "cannot connect." All of these errors mean the same thing. The local system has a route to the remote system, but it receives no response from the remote system to any of the packets it sends. There are many possible causes of this problem. The remote host may be down. Either the local or the remote host may be configured incorrectly. A gateway or circuit between the local host and the remote host may be down. The remote host may have routing problems. Only additional testing can isolate the cause of the problem. Carefully check the local configuration using netstat and ifconfig. Check the route to the remote system with traceroute. Contact the administrator of the remote system and report the problem.

Diagnostic Tools Summary

All of the tools mentioned here will be discussed later in this chapter. However, before leaving ping, let's look more closely at the command and the statistics it displays.

Troubleshooting Network Access. The "no answer" and "cannot connect" errors indicate a problem in the lower layers of the network protocols. If the preliminary tests point to this type of problem, concentrate your testing on routing and on the network interface. Use the ifconfig, netstat, and arp commands to test the network access layer.

Troubleshooting with the ifconfig Command

Ifconfig checks the network interface configuration. Use this command to verify the user's configuration if the user's system has been recently configured or if the user's system cannot reach the remote host while other systems on the same network can.

An incorrectly set IP address can be a subtle problem. If the network part of the address is incorrect, every ping will fail with the "no answer" error. In this case, using ifconfig will reveal the incorrect address. However, if the host part of the address is wrong, the problem can be more difficult to detect. A small system, such as a PC that only connects out to other systems and never accepts incoming connections, can run for a long time with the wrong address without its user noticing the problem. Additionally, the system suffering the ill effects may not be the one that is misconfigured. It is possible for someone to accidentally use your IP address on her system, and for her mistake to cause your system intermittent communications problems. An example of this problem is discussed later in the book This type of configuration error cannot be discovered by ifconfig because the error is on a remote host. The arp command is used for this type of problem.

Network Hardware Problems

Some of the tests discussed in this section can detect a network hardware problem. If a hardware problem is indicated, contact the people responsible for the hardware. If the problem appears to be in a leased telephone line, contact the telephone company. If the problem appears to be in a WAN, contact the management of that network. Don't sit on a problem expecting it to go away. It could easily get worse.

If the problem is in your LAN, you will have to handle it yourself. Some tools, such as the cable tester described earlier, can help. But frequently the only way to approach a hardware problem is with brute force—disconnecting pieces of hardware until you find the piece causing the problem. It is most convenient to do this at the switch or hub. If you identify a device causing the problem, repair or replace it. Remember that the problem can be the cable itself, rather than any particular device.

CHECKING ROUTING

The "network unreachable" error message clearly indicates a routing problem. If the problem is in the local host's routing table, it is easy to detect and resolve. First, use netstat -nr and grep to see whether or not a valid route to your destination is installed in the routing table. This example checks for a specific route to network 128.8.0.0.

```
% netstat -nr | grep '128\.8\.0'
128.8.0.0      26.20.0.16      UG      0     37     std0
```

If this same test were run on a system that did not have this route in its routing table, it would return no response at all. For example, a user reports that the network is down because he cannot FTP to sunsite.unc.edu, and a ping test returns the following results:

```
% ping -s sunsite.unc.edu 56 2

PING sunsite.unc.edu: 56 data bytes

sendto: Network is unreachable

ping: wrote sunsite.unc.edu 64 chars, ret=-1

sendto: Network is unreachable

ping: wrote sunsite.unc.edu 64 chars, ret=-1

—sunsite.unc.edu PING Statistics—

2 packets transmitted, 0 packets received, 100% packet loss
```

Based on the "network unreachable" error message, check the user's routing table. In our example, we're looking for a route to sunsite.unc.edu. The IP address of sunsite.unc.edu is 152.2.254.81, which is a class-B address. Remember that routes are network oriented. So we check for a route to network 152.2.0.0:

CHAPTER 8: GUIDE TO MANAGING A SMALL BUSINESS LAN

Use nslookup to find the IP address if you don't know it.

```
% netstat -nr | grep '152\.2\.0\.0'
%
```

This test shows that there is no specific route to 152.2.0.0. If a route was found, grep would display it. Since there's no specific route to the destination, remember to look for a default route. This example shows a successful check for a default route:

```
% netstat -nr | grep def
default        172.16.12.1     UG    0    101277   le0
```

If netstat shows the correct specific route or a valid default route, the problem is not in the routing table. In that case, use traceroute to trace the route all the way to its destination.

If netstat doesn't return the expected route, it's a local routing problem. There are two ways to approach local routing problems, depending on whether the system uses static or dynamic routing. If you're using static routing, install the missing route using the route add command. Remember, most systems that use static routing rely on a default route, so the missing route could be the default route.

If you're using dynamic routing, make sure that the routing program is running. For example, the command below makes sure that gated is running.

```
% ps `cat /etc/gated.pid`
  PID TT STAT   TIME COMMAND
27711 ?   S   304:59 gated -tep /etc/log/gated.log
```

If the correct routing daemon is not running, restart it and specify tracing. Tracing enables you to check for problems that might be causing the daemon to terminate abnormally.

SUMMARY

Every network will have problems. This chapter discusses the tools and techniques that can help you recover from these problems, as well as the planning and monitoring that can help avoid them. A solution is sometimes obvious if you can just gain enough information about the problem. UNIX provides several built-in software tools that can help you gather information about system configuration, addressing, routing, name service, and other vital network components. Gather your tools and learn how to use them before a breakdown occurs.

CHAPTER 9

How LANs Use Protocols and Layering

In this chapter, we will address some fundamental concepts and terms used in the evolving language of internetworking. In the same way that this book provides a foundation for understanding modern networking, this chapter summarizes some common themes presented throughout the remainder of the book. Topics include flow control, error checking, and multiplexing, but the main focus of this chapter is on mapping the *OSI Reference Model* (OSI Model, for short) to networking/internetworking functions and on summarizing the general nature of addressing schemes within the context of the OSI Model.

WHAT IS AN INTERNETWORK?

An internetwork is a collection of individual networks, connected by intermediate networking devices, that functions as a single large network. Internetworking refers to the industry, products, and procedures that meet the challenge of creating and administering internetworks. Figure 9.1 illustrates some different kinds of network technologies that can be interconnected by routers and other networking devices to create an internetwork.

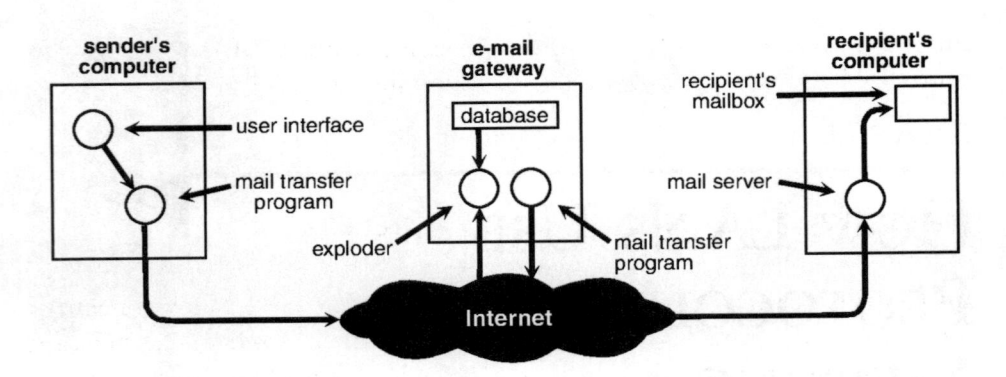

FIGURE 9.1 Different Network Technologies Connected to Create an Internetwork

HISTORY OF INTERNETWORKING

The first networks were time-sharing networks that used mainframes and attached terminals. Such environments were implemented by both IBM's SNA and Digital's network architecture.

LANs evolved around the PC revolution and enabled multiple users in a relatively small geographical area to exchange files and messages, as well as to access shared resources such as file servers.

WANs interconnect LANs across normal telephone lines (and other media), thereby interconnecting geographically dispersed users.

Today, high-speed LANs and switched internetworks are becoming widely used, largely because they operate at very high speeds and support such high-bandwidth applications as voice and videoconferencing.

Internetworking evolved as a solution to three key problems: isolated LANs, duplication of resources, and a lack of network management. Isolated LANS made electronic communication between different offices or departments impossible. Duplication of resources meant that the same hardware and software had to be supplied to each office or department, as did a separate support staff. This lack of network management meant that there was no centralized method of managing and troubleshooting networks.

INTERNETWORKING CHALLENGES

Implementing a functional internetwork is no simple task. Many challenges must be faced, especially in the areas of connectivity, reliability, network management, and flexibility. Each area is key in establishing an efficient and effective internetwork.

The challenge when connecting various systems is to support communication between disparate technologies. Different sites, for example, may use different types of media, or they might operate at varying speeds.

Reliable service, another essential consideration, must be maintained in any internetwork. Individual users and entire organizations depend on consistent, reliable access to network resources.

Furthermore, network management must provide centralized support and troubleshooting capabilities in an internetwork. Configuration, security, performance, and other issues must be adequately addressed for the internetwork to function smoothly.

Flexibility, the final concern, is necessary for network expansion and new applications and services, among other factors.

OPEN SYSTEMS INTERCONNECTION REFERENCE MODEL

The OSI Model describes how information from a software application in one computer moves through a network medium to a software application in another computer. The OSI Model is a conceptual model composed of seven layers, each specifying particular network functions. The model was developed by the International Organization for Standardization (ISO) in 1984, and it is now considered the primary architectural model for intercomputer communications. The OSI Model divides the tasks involved with moving information between networked computers into seven smaller, more manageable task groups. A task or group of tasks is then assigned to each of the seven OSI layers. Each layer is reasonably self-contained so that the tasks assigned to each layer can be implemented independently. This enables the solutions offered by one layer to be updated without adversely affecting the other layers. The following list details the seven layers of the OSIModel.

Layer 1—physical layer

Layer 2—data link layer

Layer 3—network layer

Layer 4—transport layer

Layer 5—session layer

Layer 6—presentation layer

Layer 7—application layer

Figure 9.2 illustrates the seven-layer OSI Model.

Characteristics of the OSI Layers

The seven layers of the OSI Model can be divided into two categories: application services (the *upper layers*) and networking (the *lower layers*). Figure 9-2 shows the division between the upper and lower OSI layers.

The *upper layers* of the OSI Model deal with application issues and generally are implemented only in software. The highest layer—application—is closest to the end user. Both users and application-layer processes interact with software applications that contain a communications component. The term *upper layer* is sometimes used to refer to any layer above another layer in the OSI Model.

The *lower layers* of the OSI Model handle data transport issues. The physical layer and data-link layer are implemented in hardware and software. The other lower layers generally are implemented only in software. The lowest

Application Services	**Layer 7 (Application)** - Communications-related services oriented towards specific applications. Examples include file transfer and email.
	Layer 6 (Presentation) - Negotiates formats, transforms information into agreed-upon format, generates session requests for service
	Layer 5 (Session) - Manages connections between cooperating applications by establishing and releasing sessions, synchronizing information transfer over these sessions, mapping session-to-transport and session-to-application sessions.
Networking	**Layer 4 (Transport)** - Manages connections between two end nodes by establishing and releasing end-to-end connections; controlling the size, sequence, and flow of transport packets; mapping transport and network addresses.
	Layer 3 (Network) - Routes information among source, intermediate, and destination nodes; establishes and maintains connections, if using connection-oriented exchanges or protocols.
	Layer 2 (Data Link) - Transfers data frames over the physical layer; responsible for reliability.
Transmission	**Layer 1 (Physical)** - Mechanical, electrical, functional, and procedural aspects of data circuits among network nodes.

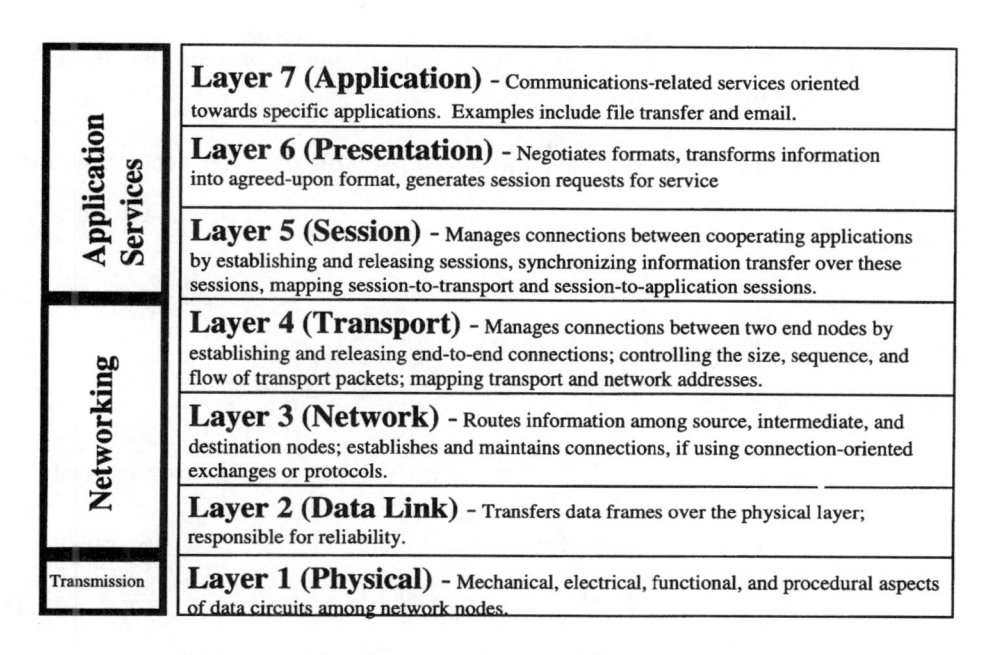

FIGURE 9.2 The OSI Model and Its Seven Independent Layers

layer—the physical layer—is closest to the physical network medium (the network cabling, for example) and is responsible for actually placing information on the medium.

The OSI Model provides a conceptual framework for communication between computers, but the model itself is not a method of communication. Actual communication is made possible by using communication protocols. In the context of data networking, a *protocol* is a formal set of rules and conventions that governs how computers exchange information over a network medium. A protocol implements the functions of one or more of the OSI layers. A wide variety of communication protocols exist, but all tend to fall into one of the following groups:

1. *LAN protocols* operate at the physical and data-link layers of the OSI Model and define communication over the various LAN media.

2. *WAN protocols* operate at the lowest three layers of the OSI Model and define communication over the various wide-area media.

3. *Network protocols* are the various upper-layer protocols that exist in a given protocol suite.

4. *Routing protocols* are network-layer protocols that are responsible for path determination and traffic switching.

OSI MODEL AND COMMUNICATION BETWEEN SYSTEMS

Information being transferred from a software application in one computer system to a software application in another must pass through each of the OSI layers. For example, if a software application in System A has information to transmit to a software application in System B, the application program in System A will pass its information to the application layer (layer 7) of System A. The application layer then passes the information to the presentation layer (layer 6), which relays the data to the session layer (layer 5), and so on down to the physical layer (layer 1). At the physical layer, the information is placed on the physical network medium and is sent across the medium to System B. The physical layer of System B removes the information from the physical medium, and then its physical layer passes the information up to the data-link layer (layer 2), which passes it to the network layer (layer 3), and so on, until it reaches the application layer (layer 7) of System B. Finally, the application layer of System B passes the information to the recipient application program to complete the communication process.

INTERACTION BETWEEN OSI MODEL LAYERS

A given layer in the OSI Model generally communicates with three other OSI layers: the layer directly above it, the layer directly below it, and its peer layer in other networked computer systems. The data-link layer in System A, for example, communicates with the network layer of System A, the physical layer of System A, and the data-link layer in System B. Figure 9.3 illustrates this example.

Data Flow on a Single Network

| Application Layer |
| Presentation Layer |
| Session Layer |
| Transport Layer |
| Network Layer |
| Data Link Layer |
| Physical Layer |

| Application Layer |
| Presentation Layer |
| Session Layer |
| Transport Layer |
| Network Layer |
| Data Link Layer |
| Physical Layer |

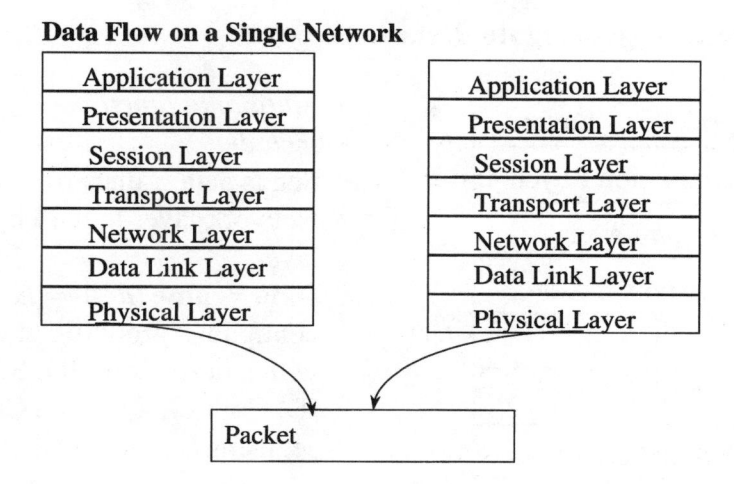

Packet

FIGURE 9.3 OSI Model Layers Communicating with Other Layers

OSI-Layer Services

One OSI layer communicates with another layer to make use of the services provided by the second layer. The services provided by adjacent layers help a given OSI layer communicate with its peer layer in other computer systems. Three basic elements are involved in layer services: the service user, the service provider, and the SAP. In this context, the service *user* is the OSI layer that requests services from an adjacent OSI layer. The service *provider* is the OSI layer that provides services to service users. OSI layers can provide services to multiple service users. The *SAP* is a conceptual location at which one OSI layer can request the services of another OSI layer.

Figure 9.4 illustrates how these three elements interact at the network and data-link layers.

OSI Model Layers and Information Exchange

The seven OSI layers use various forms of *control information* to communicate with their peer layers in other computer systems. This *control information* consists of specific requests and instructions that are exchanged between peer OSI layers.

More on Exploring the Data Link Layer

Application Layer
Presentation Layer
Session Layer
Transport Layer
Network Layer
Data Link Layer
Physical Layer

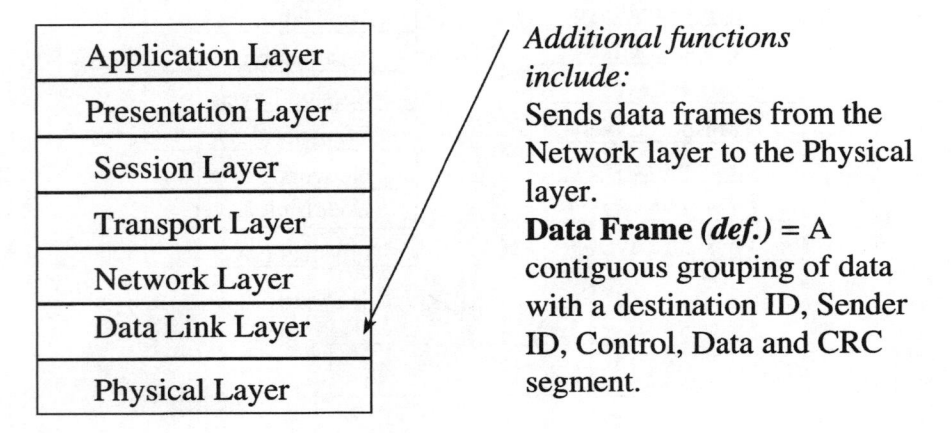

Additional functions include:
Sends data frames from the Network layer to the Physical layer.
Data Frame *(def.)* = A contiguous grouping of data with a destination ID, Sender ID, Control, Data and CRC segment.

FIGURE 9.4 Service Users, Providers and ASPs interact at the Network and Data Link layers

Control information typically takes one of two forms: headers and trailers. Headers are prefixed to data that has been passed down from upper layers. Trailers are appended to data that has been passed down from upper layers. An OSI layer is not required to attach a header or trailer to data from upper layers.

Headers, trailers, and data are relative concepts, depending on the layer that analyzes the information unit. At the network layer, an information unit, for example, consists of a layer-3 header and data. At the data-link layer, however, all the information passed down by the network layer (the layer-3 header and the data) is treated as data.

In other words, the data portion of an information unit at a given OSI layer can potentially contain headers, trailers, and data from all the higher layers. This is known as *encapsulation*. Figure 9.5 shows how the header and data from one layer are encapsulated into the header of the next lowest layer.

INFORMATION EXCHANGE PROCESS

The information exchange process occurs between peer OSI layers. Each layer in the source system adds control information to data, and each layer

- **Data Frames - A Closer Look**
  - What a Data Frame's Structure looks like

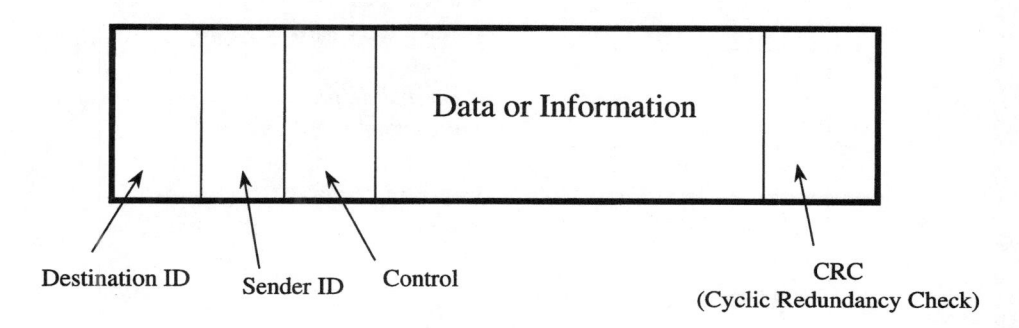

Destination ID | Sender ID | Control | Data or Information | CRC (Cyclic Redundancy Check)

FIGURE 9.5 Headers and Data Encapsulated during Information Exchange

in the destination system analyzes and removes the control information from that data.

If System A has data from a software application to send to System B, the data is passed to the application layer. The application layer in System A then communicates any control information required by the application layer in System B by prefixing a header to the data. The resulting information unit (a header and the data) is passed to the presentation layer, which prefixs its own header containing control information intended for the presentation layer in System B. The information unit grows in size as each layer prefixs its own header (and in some cases a trailer) that contains control information to be used by its peer layer in System B. At the physical layer, the entire information unit is placed onto the network medium.

The physical layer in System B receives the information unit and passes it to the data-link layer. The data-link layer in System B then reads the control information contained in the header prefixed by the data-link layer in System A. The header is then removed, and the remainder of the information unit is passed to the network layer. Each layer performs the same actions: The layer reads the header from its peer layer, strips it off, and passes the remaining information unit to the next highest layer. After

Exploring the Physical Layer

Application Layer
Presentation Layer
Session Layer
Transport Layer
Network Layer
Data Link Layer
Physical Layer

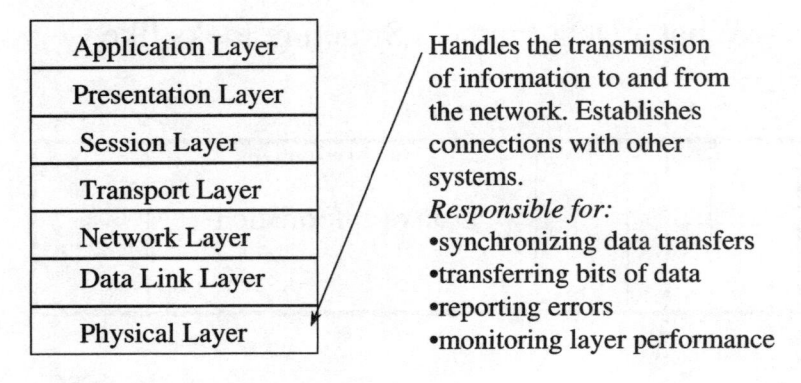

Handles the transmission of information to and from the network. Establishes connections with other systems.
Responsible for:
- synchronizing data transfers
- transferring bits of data
- reporting errors
- monitoring layer performance

FIGURE 9.6 Physical-Layer Implementations—LAN or WAN Specifications

the application layer performs these actions, the data is passed to the recipient software application in System B, in exactly the form in which it was transmitted by the application in System A.

PHYSICAL LAYER

The physical layer defines the electrical, mechanical, procedural, and functional specifications for activating, maintaining, and deactivating the physical link between communicating network systems. Physical-layer specifications define characteristics such as voltage levels, timing of voltage changes, physical data rates, maximum transmission distances, and physical connectors. Physical-layer implementations can be categorized as either LAN or WAN specifications. Figure 9.6 illustrates some common LAN and WAN physical-layer implementations.

DATA-LINK LAYER

The data-link layer provides reliable transit of data across a physical network link. Different data-link-layer specifications define different network and protocol characteristics, including physical addressing, network topology,

More on Exploring the Data Link Layer

Application Layer
Presentation Layer
Session Layer
Transport Layer
Network Layer
Data Link Layer
Physical Layer

Data Link layer is responsible for moving data frames from one computer to another through the physical layer. When data frames are sent from one computer to another it is the Data Link layer that waits for acknowledgment. Frames not acknowledged are resent.

FIGURE 9.7 Data Link Layer and Two Sublayers

error notification, sequencing of frames, and flow control. Physical addressing (as opposed to network addressing) defines how devices are addressed at the data-link layer. Network topology consists of the data-link-layer specifications that often define how devices are to be physically connected, such as in a bus or a ring topology. Error notification alerts upper-layer protocols that a transmission error has occurred, and the sequencing of data frames reorders frames that are transmitted out of sequence. Finally, flow control moderates the transmission of data so that the receiving device is not overwhelmed with more traffic than it can handle at one time.

IEEE has subdivided the data-link layer into two sublayers: Logical Link Control (LLC) and MAC. Figure 9.7 illustrates the IEEE sublayers of the data-link layer.

The LLC sublayer of the data-link layer manages communications between devices over a single link of a network. LLC is defined in the IEEE 802.2 specification and supports both connectionless and connection-oriented services used by higher-layer protocols. IEEE 802.2 defines a number of fields in data–link-layer frames that enable multiple higher-layer protocols to share a single physical data link. The MAC sublayer of the data-link layer manages protocol access to the physical network medium. The IEEE MAC specification defines MAC addresses, which enable multiple devices to uniquely identify one another at the data link layer.

Network Layer

The network layer provides routing and related functions that enable multiple data links to be combined into an internetwork. This is accomplished by the logical addressing (as opposed to the physical addressing) of devices. The network layer supports both connection-oriented and connectionless service from higher-layer protocols. Network-layer protocols typically are routing protocols, but other types of protocols are implemented at the network layer as well. Some common routing protocols include Border Gateway Protocol (BGP), an Internet interdomain routing protocol; Open Shortest Path First (OSPF), a link-state, interior gateway protocol developed for use in TCP/IP networks; and Routing Information Protocol (RIP), an Internet routing protocol that uses hop count as its metric.

Transport Layer

The transport layer implements reliable internetwork data transport services that are transparent to upper layers. Transport-layer functions typically include flow control, multiplexing, virtual circuit management, and error checking and recovery.

Flow control manages data transmission between devices so that the transmitting device does not send more data than the receiving device can process. Multiplexing enables data from several applications to be transmitted onto a single physical link. Virtual circuits are established, maintained, and terminated by the transport layer. Error checking involves creating various mechanisms for detecting transmission errors, while error recovery involves taking an action, such as requesting that data be retransmitted, to resolve any errors that occur.

Some transport-layer implementations include TCP, Name Binding Protocol (NBP), and OSI transport protocols. TCP is the protocol in the TCP/IP suite that provides reliable transmission of data. NBP is the protocol that associates AppleTalk names with addresses. OSI transport protocols are a series of transport protocols in the OSI protocol suite.

Session Layer

The session layer establishes, manages, and terminates communication sessions between presentation-layer entities. Communication sessions consist of service requests and service responses that occur between applications located in different network devices. These requests and responses are coordinated by protocols implemented at the session layer. Some examples of session-layer implementations include Zone Information Protocol (ZIP), the AppleTalk protocol that coordinates the name binding process, and Session Control Protocol (SCP), the DECnet Phase IV session-layer protocol.

Presentation Layer

The presentation layer provides a variety of coding and conversion functions that are applied to application-layer data. These functions ensure that information sent from the application layer of one system will be readable by the application layer of another system. Some examples of presentation-layer coding and conversion schemes include common data representation formats, conversion of character representation formats, common data compression schemes, and common data encryption schemes.

Common data representation formats—the use of standard image, sound, and video formats—enable the interchange of application data between different types of computer systems. Conversion schemes are used to exchange information with systems by using different text and data representations, such as EBCDIC and ASCII. Standard data compression schemes enable data that is compressed at the source device to be properly decompressed at the destination. Standard data encryption schemes enable data encrypted at the source device to be properly deciphered at the destination.

Presentation-layer implementations are not typically associated with a particular protocol stack. Some well-known standards for video include QuickTime and Motion Picture Experts Group (MPEG). QuickTime is an Apple Computer specification for video and audio, and MPEG is a standard for video compression and coding.

Among the well-known graphic image formats are Graphics Interchange Format (GIF), Joint Photographic Experts Group (JPEG), and Tagged Image File Format (TIFF). GIF and JPEG are standards for compressing and coding graphic images. TIFF is a standard coding format for graphic images.

APPLICATION LAYER

The application layer is the OSI layer closest to the end user, which means that both the OSI application layer and the user interact directly with the software application.

This layer interacts with software applications that implement a communicating component. Such application programs fall outside the scope of the OSI Model. Application-layer functions typically include identifying communication partners, determining resource availability, and synchronizing communication.

When identifying communication partners, the application layer determines the identity and availability of communication partners for an application with data to transmit. When determining resource availability, the application layer must decide whether sufficient network resources exist for the requested communication. All communication between applications requires cooperation that is managed by the application layer.

Two key types of application-layer implementations are TCP/IP applications and OSI applications. TCP/IP applications are protocols, such as telnet, FTP, and Simple Mail Transfer Protocol (SMTP), that exist in the IP suite. OSI applications are protocols, such as File Transfer, Access, and Management (FTAM), Virtual Terminal Protocol (VTP), and Common Management Information Protocol (CMIP), that exist in the OSI suite.

INFORMATION FORMATS

The data and control information transmitted through internetworks takes a wide variety of forms. The terms used to refer to these information formats are not used consistently in the internetworking industry, but sometimes they are used interchangeably. Common information formats include frame, packet, datagram, segment, message, cell, and data unit.

A *frame* is an information unit whose source and destination are entities of the data link layer. A frame is composed of the data link–layer header (and possibly a trailer) and upper-layer data. The header and trailer contain control information intended for the data link–layer entity in the destination system. Data from upper-layer entities is encapsulated in the data link–layer header and trailer.

A *packet* is an information unit whose source and destination are network-layer entities. A packet is composed of the network-layer header (and possibly a trailer) and upper-layer data. The header and trailer contain control information intended for the network-layer entity in the destination system. Data from upper-layer entities is encapsulated in the network-layer header and trailer. Figure 9.8 illustrates the basic components of a data frame.

The term *datagram* usually refers to an information unit whose source and destination are network-layer entities that use connectionless network service.

- **Data Frames - A Closer Look**
  - What a Data Frame's Structure looks like

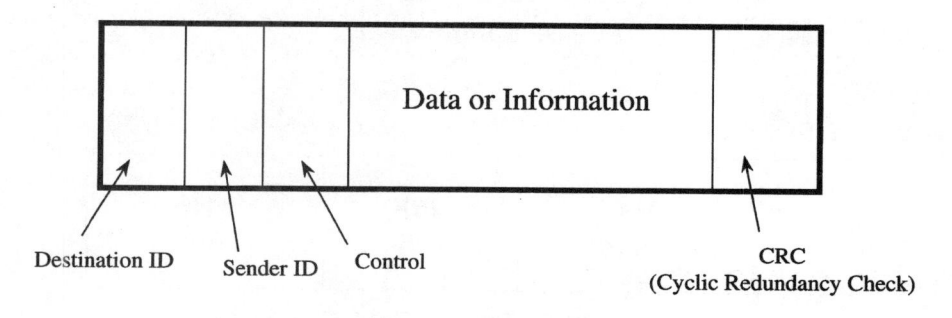

Destination ID | Sender ID | Control | Data or Information | CRC (Cyclic Redundancy Check)

FIGURE 9.8 Three Basic Components that Make Up a Network-Layer Packet

The term *segment* usually refers to an information unit whose source and destination are transport-layer entities.

A *message* is an information unit whose source and destination entities exist above the network layer (often at the application layer).

A *cell* is an information unit of a fixed size whose source and destination are data-link–layer entities. Cells are used in switched environments such as ATM and Switched Multimegabit Data Service (SMDS) networks. A cell is composed of the header and payload. The header contains control information intended for the destination data-link–layer entity and is typically 5 bytes long. The payload contains upper-layer data that is encapsulated in the cell header and is typically 48 bytes long. The length of the header and the payload fields are always exactly the same for each cell. Figure 9.9 depicts the components of a typical cell.

Data unit is a generic term that refers to a variety of information units. Some common data units are service data units (SDUs), protocol data units (PDUs), and bridge protocol data units (BPDUs). SDUs are information

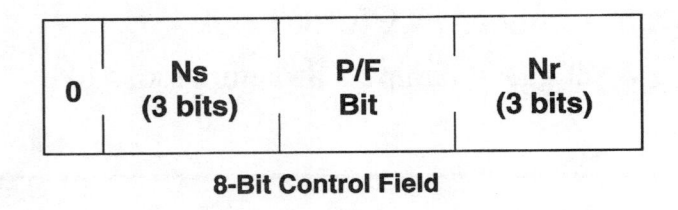

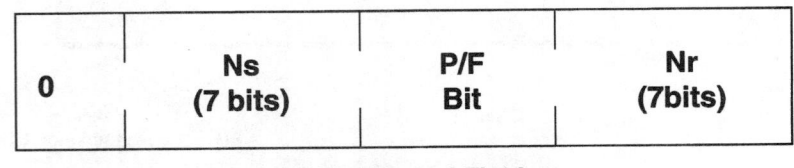

0	Ns (3 bits)	P/F Bit	Nr (3 bits)

8-Bit Control Field

0	Ns (7 bits)	P/F Bit	Nr (7bits)

16-Bit Control Field

FIGURE 9.9 Two Components that Make Up a Typical Cell

units from upper-layer protocols that define a service request to a lower-layer protocol. PDU is OSI terminology for a packet. BPDUs are used by the spanning-tree algorithm as hello messages.

ISO HIERARCHY OF NETWORKS

Large networks typically are organized as hierarchies. A hierarchical organization provides such advantages as ease of management, flexibility, and a reduction in unnecessary traffic. Thus, the ISO has adopted a number of terminology conventions for addressing network entities. Key terms, defined in this section, include *end system* (ES), *intermediate system* (IS), *area*, and *autonomous system* (AS). An *ES* is a network device that does not perform routing or other traffic-forwarding functions. Typical ESs include such devices as terminals, PCs, and printers. An *IS* is a network device that performs routing or other traffic-forwarding functions. Typical *ISs* include such devices as routers, switches, and bridges. Two types of IS networks exist: intradomain IS and interdomain IS. An intradomain IS communicates within a single autonomous system, while an interdomain IS communicates within and between autonomous systems. An *area* is a logical group of network segments and their attached devices. Areas are subdivisions of ASs. An AS is a collection of networks under a common administration that share a common routing strategy. Autonomous systems are subdivided into areas, and an AS is sometimes called a *domain*. Figure 9.10 illustrates a hierarchical network and its components.

CONNECTION-ORIENTED AND CONNECTIONLESS NETWORK SERVICES

In general, networking protocols and the data traffic they support can be characterized as being either connection oriented or connectionless. In brief, connection-oriented data handling involves using a specific path that is established only for the duration of a connection. Connectionless data handling involves passing data through a permanently established connection.

Hierarchical Network Topology

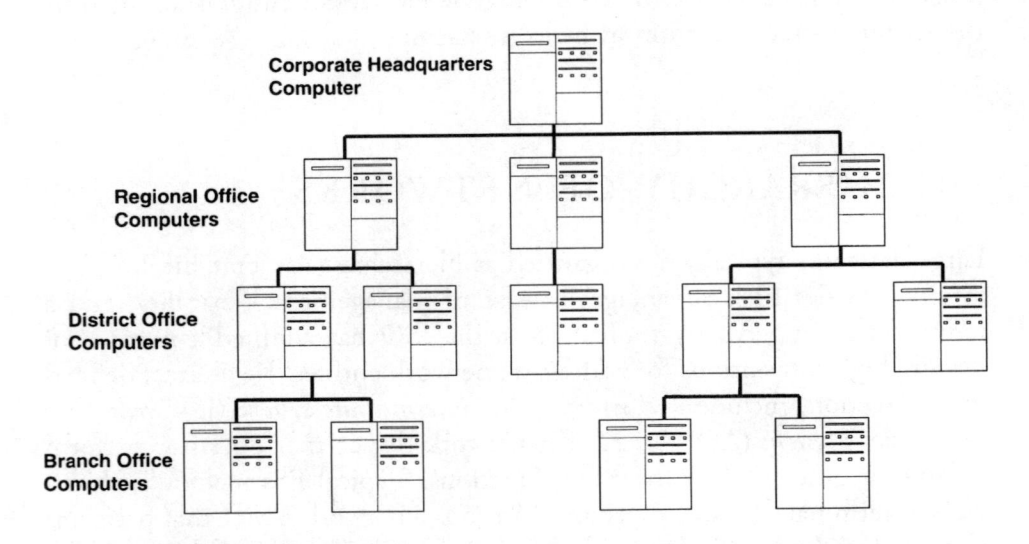

FIGURE 9.10 A Hierarchical Network and Its Components

CONNECTION-ORIENTED SERVICES

Connection-oriented service involves three phases:

1. Connection establishment

2. Data transfer

3. Connection termination

During the connection-establishment phase, a single path is determined between the source and destination systems. Network resources typically are reserved at this time to ensure a consistent grade of service, such as a guaranteed throughput rate.

In the data-transfer phase, data is transmitted sequentially over the path that has been established. Data always arrives at the destination system in the order in which it was sent.

During the connection-termination phase, an established connection that is no longer needed is terminated. If there is to be further communication between the source and destination systems, a new connection must be established.

Disadvantages

Connection-oriented network service carries two significant disadvantages over connectionless: static-path selection and the static reservation of network resources. Static-path selection can create difficulty because all traffic must travel along the same static path. A failure anywhere along that path causes the connection to fail. Static reservation of network resources causes difficulty because it requires a guaranteed rate of throughput and, thus, a commitment of resources that other network users cannot share. Unless the connection uses full, uninterrupted throughput, bandwidth is not used efficiently. Another disadvantage of connection-oriented network service is that it does not predetermine the path from the source to the destination system, nor are packet sequencing, data throughput, and other network resources guaranteed. Each packet must be completely addressed because different paths through the network may be selected for different packets, based on a variety of influences. Each packet is transmitted independently by the source system and is handled independently by intermediate network devices.

Advantages

On the positive side, however, connection-oriented services are useful for transmitting data from applications that don't tolerate delays and packet resequencing. Voice and video applications are typically based on connection-oriented services.

CONNECTIONLESS SERVICES

As mentioned above, connectionless data handling involves passing data through a permanently established connection. Connectionless service offers two important advantages over connection-oriented service: dynamic-path selection and dynamic-bandwidth allocation. Dynamic-path selection enables traffic to be routed around network failures because paths are selected on a packet-by-packet basis. With dynamic-bandwidth allocation, bandwidth is used more efficiently because bandwidths are not allocated to network resources that the resources will not use.

Connectionless services are useful for transmitting data from applications that can tolerate some delay and resequencing. Data-based applications typically are based on connectionless service.

INTERNETWORK ADDRESSING

Internetwork addresses identify devices separately or as members of a group. Addressing schemes vary, depending on the protocol family and the OSI layer. Three types of internetwork addresses are commonly used: data-link–layer addresses, MAC addresses, and network-layer addresses.

DATA-LINK–LAYER ADDRESSES

A data-link-layer address uniquely identifies each physical network connection of a network device. Data-link–layer addresses are sometimes referred to as *physical* or *hardware* addresses. Data-link–layer addresses usually exist within a flat address space and have a preestablished and typically fixed relationship to a specific device.

End systems generally have only one physical network connection and thus have only one data-link–layer address. Routers and other internetworking devices typically have multiple physical network connections and therefore have multiple data link addresses.

MAC ADDRESSES

MAC addresses consist of a subset of data-link–layer addresses. MAC addresses identify network entities in LANs that implement the IEEE MAC

IP and IPX Header Formats

Version	IP Header Length	Type of Service	Packet Length	Identification	Flags	Fragment Offset	Time to Live	Protocol	Header Checksum	Source Address	Destination Address

IP Header

Checksum	Length	Transport Control	Packet Type	Destination Network	Destination Node	Destination Socket	Source Network	Source Node	Source Socket

IPX Header

FIGURE 9.11 MAC Address with Unique Format of Hexadecimal Digits

addresses of the data-link layer. As with most data-link–layer addresses, MAC addresses are unique for each LAN interface.

MAC addresses are 48 bits in length and are expressed as 12 hexadecimal digits. The first 6 hexadecimal digits, which are administered by the IEEE, identify the manufacturer or vendor and thus comprise the Organizational Unique Identifier (OUI). The last 6 hexadecimal digits comprise the interface serial number or another value administered by the specific vendor. MAC addresses sometimes are called burned-in addresses (BIAs) because they are burned into read-only memory (ROM) and are copied into random-access memory (RAM) when the interface card initializes. Figure 9.11 illustrates the MAC address format.

Different protocol suites use different methods for determining the MAC address of a device. The two most often used are are these:

- ARP maps network addresses to MAC addresses.

- Hello protocol enables network devices to learn the MAC addresses of other network devices.

MAC addresses are either embedded in the network-layer address or generated by an algorithm. Address resolution is the process of mapping network addresses to MAC addresses. This process is accomplished by using ARP, which is implemented by many protocol suites. When a network address is successfully associated with a MAC address, the network device stores the information in the ARP cache. The ARP cache enables devices to send traffic to a destination without creating ARP traffic because the MAC address of the destination is already known.

The process of address resolution differs slightly, depending on the network environment. Address resolution on a single LAN begins when End System A broadcasts an ARP request onto the LAN in an attempt to learn the MAC address of End System B. The broadcast is received and processed by all devices on the LAN, although only End System B replies to the ARP request—it replies by sending an ARP reply containing its MAC address to End System A. End System A receives the reply and saves the MAC address of End System B in its ARP cache. (The ARP cache is where network addresses are paired with MAC addresses.) Whenever End System A must communicate with End System B, it checks the ARP cache, finds the MAC address of System B, and sends the frame directly without first having to use an ARP request.

However, address resolution works differently when the source and destination devices are attached to different LANs that are interconnected by a router. End System Y broadcasts an ARP request onto the LAN in an attempt to learn the MAC address of End System Z. The broadcast is received and processed by all devices on the LAN, including Router X, which, acting as a proxy for End System Z, checks its routing table to determine that End System Z is located on a different LAN. Router X then replies to the ARP request from End System Y, sending an ARP reply containing its *own* MAC address as if it belonged to End System Z. End System Y receives the ARP reply and saves the MAC address of Router X in its ARP cache in the entry for End System Z. When End System Y must communicate with End System Z, it checks the ARP cache, finds the MAC address of Router X, and sends the frame directly without using ARP requests. Router X receives the traffic from End System Y and forwards it to End System Z on the other LAN.

The Hello protocol is a network-layer protocol that enables network devices to identify one another and indicate that they are still functional.

When a new end system powers up, for example, it broadcasts Hello messages onto the network. Devices on the network then return Hello replies. Hello messages are also sent at specific intervals to indicate that they are still functional. Network devices can learn the MAC addresses of other devices by examining Hello protocol packets.

Three protocols use predictable MAC addresses: Xerox Network Systems (XNS), Novell Internetwork Packet Exchange (IPX), and DECnet Phase IV. In these protocol suites, MAC addresses are predictable because the network layer either embeds the MAC address in the network-layer address or uses an algorithm to determine the MAC address.

NETWORK-LAYER ADDRESSES

A network-layer address identifies an entity at the network layer of the OSI layers. Network addresses usually exist within a hierarchical address space and sometimes are called *virtual* or *logical* addresses.

The relationship between a network address and a device is logical and unfixed; typically it is based either on physical network characteristics (the device on a particular network segment) or on groupings that have no physical basis (the device being part of an AppleTalk zone). End systems require one network-layer address for each network-layer protocol they support. (This assumes that the device has only one physical network connection.) Routers and other internetworking devices require one network-layer address per physical network connection for each network-layer protocol supported. For example, a router with three interfaces—each running AppleTalk, TCP/IP, and OSI—must have three network-layer addresses for each interface. The router therefore has nine network-layer addresses.

HIERARCHICAL VERSUS FLAT ADDRESS SPACE

Internetwork address space typically takes one of two forms: hierarchical address space or flat address space. A hierarchical address space is organized into numerous subgroups, each successively narrowing an address until it points to a single device (in a manner similar to street addresses). A flat address space is organized into a single group (in a manner similar to U.S. Social Security numbers).

Hierarchical addressing offers certain advantages over flat addressing schemes. Address sorting and recall is simplified through the use of comparison operations. Ireland, for example, eliminates any other country as a possible location in its street addresses. Figure 9.12 illustrates the difference between hierarchical and flat address spaces.

Address Assignments

One of three types of addresses are assigned to devices: *static, dynamic,* or *server.*

- *Static addresses* are assigned by a network administrator according to a preconceived internetwork addressing plan. A static address does not change until the network administrator manually changes it.

- *Dynamic addresses* are obtained by devices by means of some protocol-specific process when they attach to a network. A device using a dynamic address often has a different address each time it connects to the network.

- *Server address,* or addresses that are assigned by a server, are given to devices as they connect to the network. Server-assigned addresses are recycled for reuse as devices disconnect, so a device is therefore likely to have a different address each time it connects to the network.

Addresses Versus Names

Internetwork devices usually have both a name and an address associated with them. Internetwork names typically are location independent and remain associated with a device wherever that device moves (for example, from one building to another). Internetwork addresses are usually location dependent and change when a device is moved (although MAC addresses

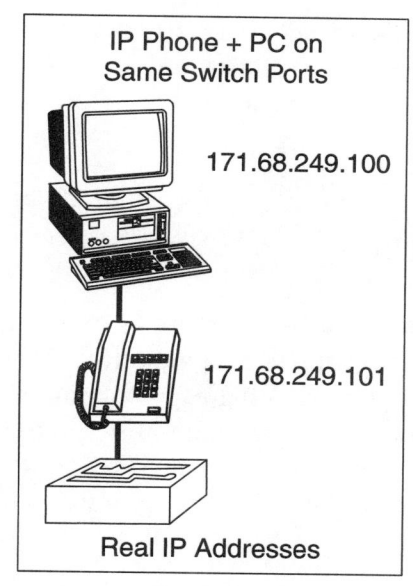

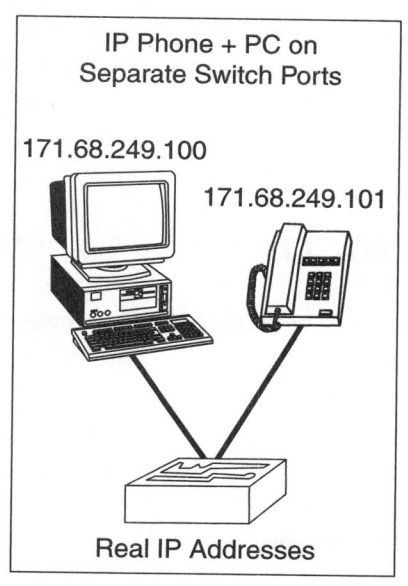

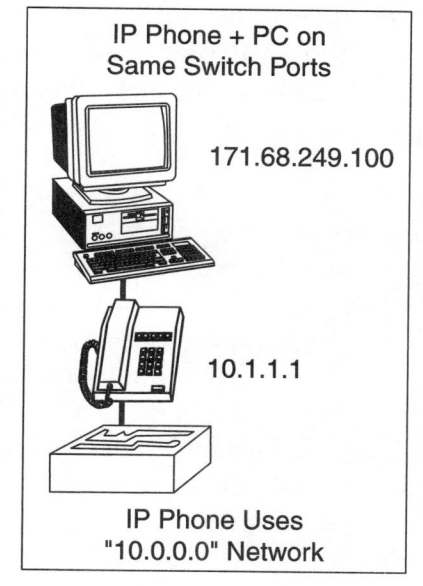

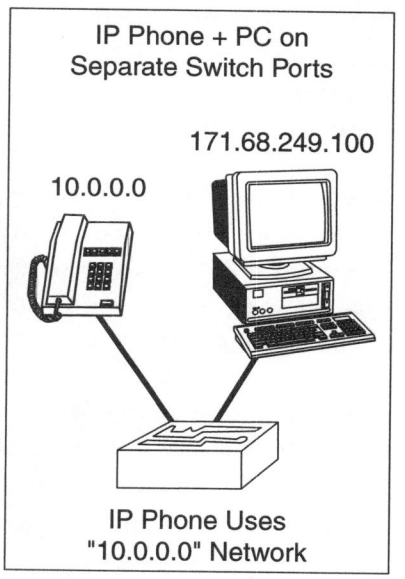

FIGURE 9.12 Hierarchical and Flat Address Spaces

are an exception to this rule). Names and addresses represent a logical identifier, which may be a local system administrator or an organization, such as the Internet Assigned Numbers Authority (IANA).

FLOW-CONTROL BASICS

Flow control is a function that prevents network congestion by ensuring that transmitting devices do not overwhelm receiving devices with data. There are countless possible causes of network congestion. A high-speed computer, for example, may generate traffic faster than the network can transfer it, or faster than the destination device can receive and process it. The three commonly used methods for handling network congestion are buffering, source-quench message transmitting, and windowing:

Buffering is used by network devices to store bursts of excess data in memory temporarily until they can be processed. Occasional data bursts are easily handled by buffering, but excess data bursts can exhaust memory, forcing the device to discard any additional datagrams that arrive.

Source-quench messages are used by receiving devices to help prevent their buffers from overflowing. The receiving device sends source-quench messages to request that the source reduce its current rate of data transmission. First, the receiving device begins discarding received data due to overflowing buffers. Second, the receiving device begins sending source-quench messages to the transmitting device at the rate of one message for each packet dropped. The source device receives the source-quench messages and lowers the data rate until it finally stops receiving the messages. Then the source device gradually increases the data rate as long as it receives no further source-quench messages.

Windowing is a flow-control scheme in which the source device requires an acknowledgment from the destination after a certain number of packets have been transmitted. With a window size of three, the source requires an acknowledgment after sending three packets, as follows: First, the source device sends three packets to the destination device. Then, after receiving the three packets, the destination device sends an acknowledgment to the source. The source receives the acknowledgment and sends three more packets. If the destination does not receive one or more of the packets for some reason, such as overflowing buffers, it does not receive the required three packets that prompt it to send an acknowledgment. The source, not receiving an acknowledgment, then retransmits the packets at a reduced transmission rate.

ERROR-CHECKING BASICS

Error-checking schemes determine whether transmitted data has become corrupt or otherwise damaged while traveling from the source to the destination. Error checking is implemented at a number of the OSI layers. One common error-checking scheme is the cyclic redundancy check (CRC), which detects and discards corrupted data. Error-correction functions (such as data retransmission) are left to higher-layer protocols. The CRC error-checking scheme proceeds as follows:

- The source device performs a predetermined set of calculations over the contents of the packet to be sent.

- The source device places the calculated value in the packet and sends the packet to the destination.

- The destination device performs the same predetermined set of calculations over the contents of the packet and then compares its computed value with that contained in the packet. If the values are equal, the destination device considers the packet to be valid. If the values are unequal, the packet contains errors and is discarded.

MULTIPLEXING BASICS

Multiplexing is a process in which multiple data channels are combined into a single data or physical channel at the source. Multiplexing can be implemented at any of the OSI layers. Conversely, demultiplexing is the process

FIGURE 9.13 Multiple Devices Multiplexed into a Single Physical Channel

of separating multiplexed data channels at the destination. One example of multiplexing occurs when data from multiple applications is multiplexed into a single lower-layer data packet.

Another example of multiplexing occurs when data from multiple devices is combined into a single physical channel using a device called a multiplexer (see Figure 9.13). A multiplexer is a physical-layer device that combines multiple data streams into one or more output channels at the source. Multiplexers demultiplex the channels into multiple data streams at the remote end and thus maximize the use of the bandwidth of the physical medium by enabling it to be shared by multiple traffic sources.

Some methods used for multiplexing data include the following:

- **Time-division multiplexing (TDM).** Information from each data channel is allocated bandwidth based on preassigned time slots, regardless of whether or not there is data to transmit.

- **Asynchronous time-division multiplexing (ATDM).** Information from data channels is allocated bandwidth as needed using dynamically assigned time slots.

- **Frequency-division multiplexing (FDM).** Information from each data channel is allocated bandwidth based on the signal frequency of the traffic.

- **Statistical multiplexing.** Bandwidth is dynamically allocated to any data channels that have information to transmit.

CHAPTER 10

INTERNETWORKING CONCEPTS FOR SMALL BUSINESS AND HOME LANS

A NEW PHASE FOR HOME NETWORKING

The world has gone bonkers over the Internet. In just a few short years, the Internet has become an indispensable part of life. Everyone is going online to send e-mail, play interactive games, chat with friends, find romance, research their roots, manage their portfolios, purchase an amazing variety of goods and services, and download information on absolutely anything under the sun.

But it's not all fun and games. More and more families include a wage earner who relies on the Internet for a home-based business, works from home one or more days each week, or travels on business and needs access to office resources and email. Students of all ages are taking Web-based courses offered through employers, local and remote universities, or online training companies. Forecasters predict that more than 30 million North American households will own two or more computers by the end of 2002 (In-Stat, 2001); those computers will need Internet access. Multiple phone lines are costly, so the ability for multiple computers to share access to the Internet using a single access line to an ISP is spurring interest in home networking. Now what people thirst for is faster access to those Internet resources. Analog modem dial-up speeds are already too slow, and they certainly won't keep up when video and audio Internet-based entertainment becomes widely available. With the increasing availability of high-speed cable and digital subscriber line (DSL) services, the power of broadband Internet access is arriving at more and more doorsteps.

HOME NETWORKING PRIMER

A network connects your computing devices—computers, printers, modems, and so on—so that multiple users can communicate with one another and share resources. With a home network, you and your family can share Internet access using one ISP account. (Note: some ISP contracts restrict Internet connections; check with your local ISP on its Internet-sharing policies.) All the computers in the house can access the same printer, modem, or other peripherals, eliminating the need to buy duplicate equipment. You can also share local files and applications or back up those files for safekeeping. With a home network, you can also play exciting multiplayer computer games, both within the home and with others on the Internet.

By all indications, the home networking outlook is rosy. After a slower-than-expected start, the market is expected to grow from 1.3 million U.S. households by end of 2000 to more than 9.5 million households by end of 2003 (The Yankee Group, 1999).

Home networks come in two basic flavors: wired and wireless. Wired networks use some form of physical cabling—telephone wires, Ethernet cabling, or indoor electrical wiring—to interconnect devices. Wireless networks don't require physical cabling; they use radio waves to send and receive signals.

WHAT IS A GATEWAY AND WHY DO I NEED ONE?

Currently there are two types of networks: computer-based networks and gateway-based networks. Computer-based networks require a networked PC or Macintosh computer equipped with special communications software. In addition to its regular computing duties, the computer takes on the job of managing network traffic both within the LAN and between the LAN in the home and the public WANs, such as the public switched telephone network (PSTN) and the Internet.

Gateway-based networks connect devices on your home network with the communications services outside your home using an always-on, dedicated, intelligent device at the boundary between the home network and the outside world. A home or residential gateway essentially acts as a translator, or bridge, between different types of networks. Gateways have been used for years in business, government, and education to connect different types of LANs—Ethernet and Token Ring, for example—as well as to translate between LANs and WANs.

Dedicated gateways handle the job of translating between networks quickly and efficiently. As shown in Figure 10.1, the home gateway collects all of the traffic from home network computers and devices such as fax machines that is destined for the Internet and forwards, or routes, that traffic through a cable or DSL modem (or backup analog modem). As you will see, gateways can interconnect a number of home-based systems that you may not think of as networks, such as home wiring or security systems.

Gateway-based networks offer a number of valuable advantages that computer-based networks just cannot provide. Gateway applications split into four dimensions that will one day converge into the integrated digital home (see Figure 10.2):

- Communication
- Productivity
- Security and home monitoring
- Entertainment

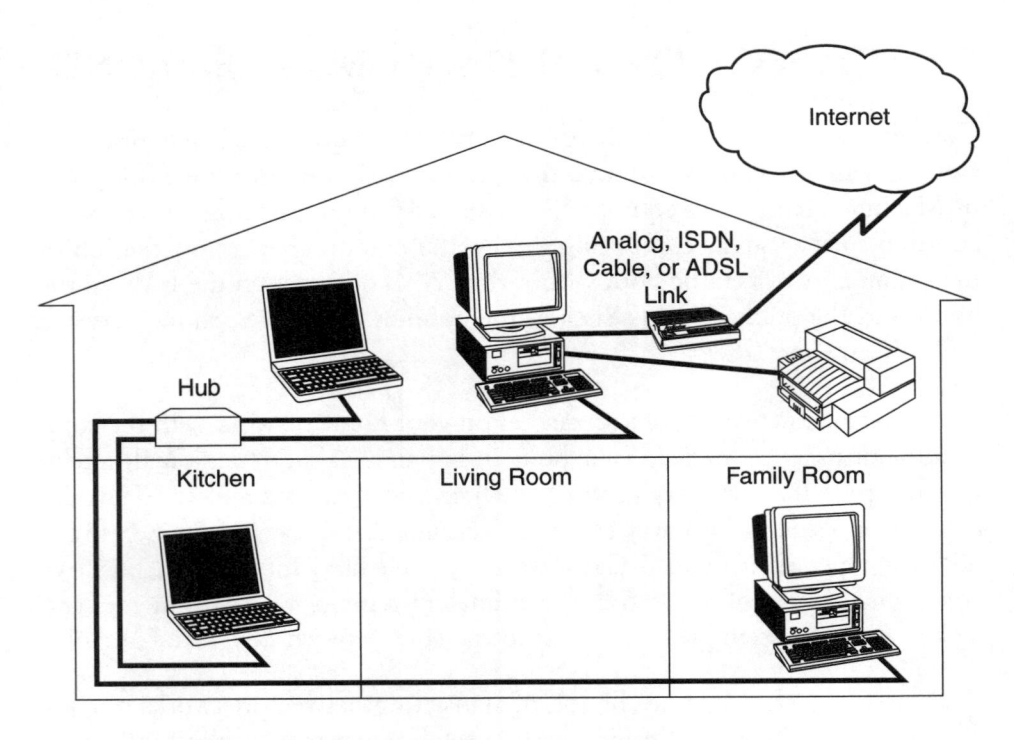

FIGURE 10.1 Gateway-Based Digital Home Network

COMMUNICATION

The primary selling point for a home gateway is the ability to share continuous (always-on), high-speed, digital Internet access. No more waiting around, listening to the screech of the phone connecting; you're on and running all the time. The gateway connects the home LAN to a WAN transport service such as DSL, cable, or satellite. Unlike computer-based networks, gateways are not tied to any particular operating system, so they can support many different types of computer systems on one network.

Gateway-based networks can also interconnect different types of network technologies, including wireless, phone line, power line, and Ethernet. This means that all types of computers and wireless PDAs can communicate on your Ethernet or phone-line network, so you now have the flexibility to purchase many types of computing and communications devices with the assurance that they will be able to communicate with one another.

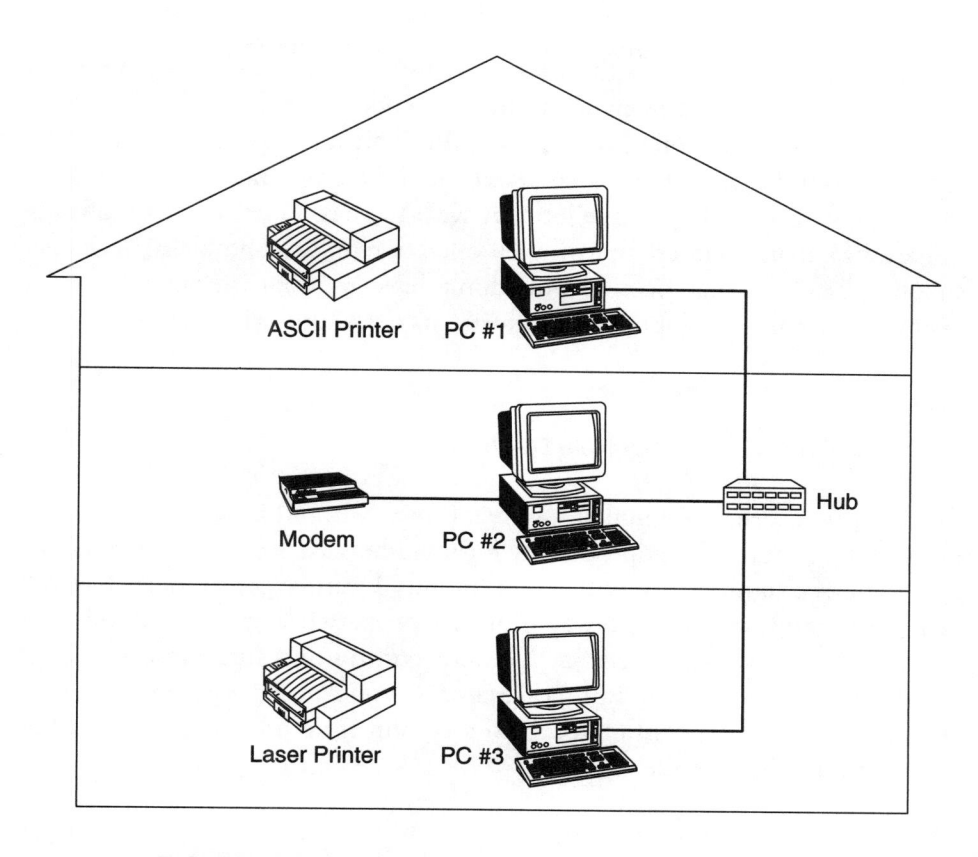

FIGURE 10.2 Dimensions of a Digital Home Network

PRODUCTIVITY

A gateway performs all traffic-handling functions for the network, enabling the networked computers to operate at their maximum efficiency. In a computer-based network, by contrast, traffic-handling functions take computer processing resources away from the computer acting as traffic manager, and this can impair overall network performance.

As new functionality such as firewall filtering is added, the productivity benefits of a dedicated gateway become even clearer. Computers can also be unreliable and difficult to configure. With a computer-based network, if the computer is turned off or not working, you and your family lose that all-important access to the Internet and to network resources.

Dedicated gateways come with software preinstalled and take very little effort to get up and running on new or existing networks. When new capabilities do become available, upgrading typically requires only a simple download of new software from the Web. Dedicated devices are inherently more reliable because they are designed to do just one job very well; there is no unnecessary software or hardware that can interfere with their operation or cause problems. Troubleshooting is also easier because you don't have to wade through layers of operating system and application software to determine what is wrong.

Security and Home Monitoring

Our increased reliance on the Internet brings with it greater concerns for privacy and security. The opportunity for unauthorized access to your personal files increases significantly with a continuous Internet connection. Firewall capabilities built into the gateway protect your network against a number of different types of hacker attacks. Content control and filtering capabilities are also built into the gateway. For example, you can limit access to certain Internet sites or files, restrict access to a certain time of day, or limit access to specific family members.

The gateway can also literally protect your home and property by acting as the control center for your home entry and lighting systems. You can set up or change settings from any computer in the home, or even from the road, through a PDA or laptop. You can also use a Webcam to monitor activities inside or outside your home from anywhere in the world.

Entertainment

In the not-too-distant future, the home network will truly encompass the entire home. The next step will incorporate home entertainment devices like the TV and video player, stereo, CD player, and radio; then it will move out to other areas of the household like the kitchen or laundry room. The availability of high-speed broadband services only accelerates the integration of the Internet inside the home, from set-top Internet appliances in the family room to screen phones in the kitchen. Next-generation dedicated home gateways will allow these seemingly different devices to communicate with one another.

The market for home gateways is expected to grow from 15,000 households by the end of 2000 to over 1.02 million households by the end of 2003, as their capabilities expand and demand increases for their communication, productivity, security, and entertainment benefits (The Yankee Group, 1999).

CHOOSING THE BEST NETWORK GATEWAY FOR YOUR HOME

No one wants to spend money on yet another network device that may add more complexity and hassle, so you must evaluate the gateway offerings on the market carefully. Several important features come into play in addition to cost. Look for these qualities in a home network gateway:

- Reliable and compatible with other network devices, as well as easy to expand as your home network grows. That means an Internet standards–based solution.

- Convenient—easy to install and set up, and easy to reconfigure when needed.

- Easy to upgrade as new capabilities and technologies become available.

- Built by reputable, experienced companies that you know and trust.

- Backed by easily accessible, professional technical support.

3COM HOMECONNECT HOME NETWORKING SOLUTIONS

With more than 20 years of networking experience, 3Com Corporation has consistently been the market leader in small business and home office networking (Dell'Oro Group, February 2000). 3Com entered the home networking market last year, introducing high-performance Ethernet, phone-line, and wireless home networking solutions that are reliable and easy to use.

3Com Ethernet, Phone-Line, and Wireless Home Networking Products

3Com HomeConnect home network products consistently offer the highest network performance that the technology allows in the three most popular home network types:

- 3Com HomeConnect phone-line products deliver 10 Mbps—10 times the performance of most competing offerings.

- 3Com HomeConnect Ethernet products support both 10-Mbps Ethernet and 100-Mbps Fast Ethernet performance for graphics-intensive applications and high-quality MPEG digital audio and video applications.

- 3Com AirConnect wireless products boast transmission rates of 11 Mbps—to 10 times the performance of most competing products.

All 3Com home networking products are designed from the same hardware and software architecture, engineered to work together seamlessly. They also all support the prevailing Internet standards, such as the HomePNA for home phone-line networking and the IEEE 802.11 for high-rate wireless networking. HomeConnect Ethernet and phone-line adapter kits come with preinstalled Microsoft HomeClick networking software that gets you up and running quickly with a simple, intuitive graphical interface. All products undergo extensive reliability and interoperability testing with a wide range of computer manufacturers. 3Com backs its HomeConnect Ethernet and phone-line home networking products with 90 days of free technical support and a five-year limited warranty.

3Com HomeConnect Cable and DSL Modems

3Com also offers a number of HomeConnect broadband cable and DSL modem products that feature a wide range of speeds, connectivity options, and prices. 3Com HomeConnect cable and DSL modems can be purchased through your telephone company, cable service provider, or authorized retail outlet.

3Com HomeConnect broadband modems are the connection point for bringing high-speed voice and data services into your home network. If you live within easy reach of cable television service, installing a cable modem can give you downstream (from the Internet to your home) speeds of up to 38 Mbps—almost 1,000 times faster than a 56 K modem! (These modems are capable of upstream speeds of up to 10 Mbps. Actual speeds will vary depending on services offered by your local broadband service provider and other factors.)

3Com HomeConnect DSL modems offer blistering speeds as well—up to 8 Mbps downstream and up to 1 Mbps upstream. (Again, actual speeds will vary depending on services offered by your local broadband service provider, the type of digital subscriber line service available, and other factors.) But the big advantage with DSL modems is their hard-to-beat flexibility. DSL modem technology transmits voice and data over the same regular phone line at the same time. That means you can be surfing the Web in the living room, for example, while someone else talks on the phone in the bedroom. Both 3Com HomeConnect broadband modems provide a continuous connection to WAN services.

3Com cable and DSL modems conform to both the Data-over-Cable Service Interface Specification (DOCSIS) standard for cable and G.dmt/G.lite standards for Asymmetric DSL (ADSL). 3Com backs its broadband cable and DSL modems with a full range of technical support services, including phone support, 24-hour automated support, and Web and email support, along with a five-year repair and replacement warranty.

New 3Com HomeConnect Gateway Offerings

3Com's initial home network gateway product offers a secure, reliable way to share continuous high-speed Internet access using a cable or DSL modem. A 56-K analog connection is also included for backup Internet access or for users who want gateway functionality now, but who are still awaiting local cable or DSL service. Preconfigured, business-quality firewall and security features protect your network and the personal information on it against unwelcome intruders.

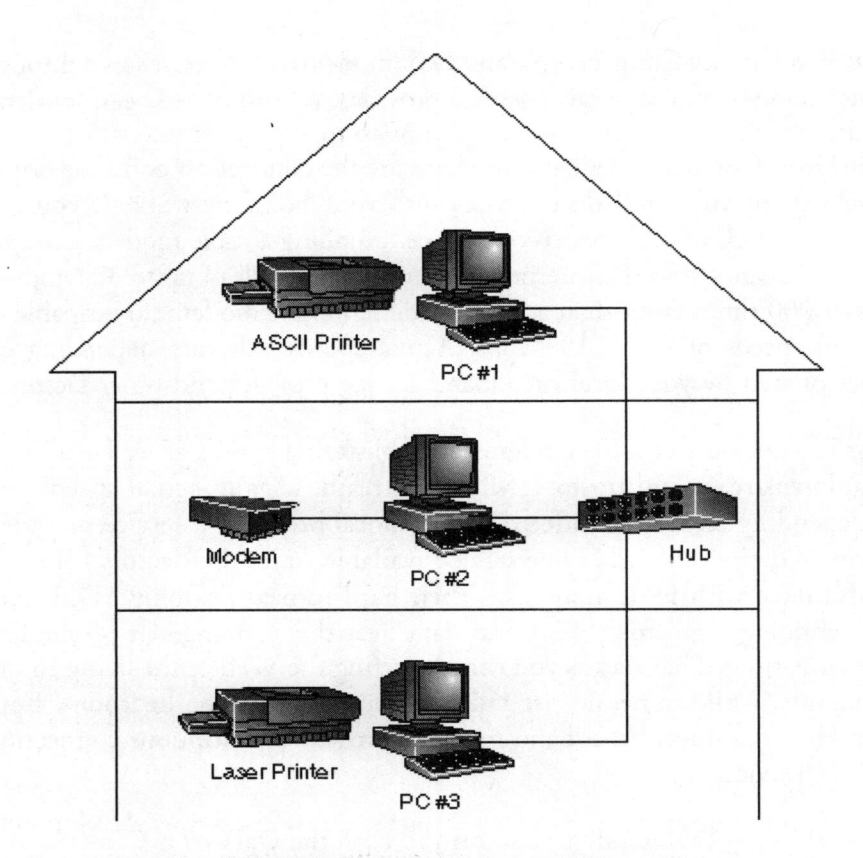

FIGURE 10.3 An HPNA home network

The flexible 3Com HomeConnect gateway features built-in Ethernet and phone-line interfaces so that computers and other devices on your Ethernet home office network can communicate with devices on your home phone-line network (see Figure 10.3)—no more network "islands." Preinstalled software cuts down on installation and setup time, while the easy-to-use Web browser interface simplifies network device management. The software also dynamically provides network settings so that you don't have to change anything as you move between home and office (not supported in all corporate environments).

Advanced features include client privileges that let you block Web, telnet, e-mail, FTP, or Network News Transfer Protocol (NNTP) access to certain computers (like the one in your teenager's room) during certain days or times. The gateway is also designed for easy upgrades so that you can download new software directly from the Web to keep your network features and functionality up to date.

SUMMARY

It's an exciting time for home networks. Multicomputer households are definitely on the rise as the power of the Web grows daily and new Internet-based applications and appliances are introduced. High-speed Internet access via DSL, cable, or satellite service is imminent if not already available in your area, unlocking the full capabilities of the Internet for home-based communications, education, commerce, entertainment, and more.

The integrated digital home will merge what we now think of as separate application dimensions—security, music and video entertainment, telephone and fax, and computing devices—into one seamless environment. The key to that future is the development of the home gateway with its ability to bridge these different systems so that they can communicate with one another. 3Com Corporation is at the forefront of home gateway development and is committed to offering the most reliable, highest-performance, most flexible, and easiest-to-use gateway products on the market.

ENDNOTES

> Cahner's In-Stat Report #RC01-06HN, "Connecting the Dots; A Complete Look at Home Network Standards and Organizations," May 2001, www.instat.com

Report, The Yankee Group, 1999, www.yankeegroup.com.

Press release, Dell'Oro Group, February 2000, www.delloro.com.

CHAPTER 11

Knowing the Fundamentals of TCP/IP for Running Your Network

An essential part of the Microsoft culture is the act of thoroughly deploying its own technology internally. Such is the case with the Microsoft implementation of TCP/IP. This approach to product designs makes the development teams accountable to their fellow employees for the performance of applications; furthermore, the development team's reputation is typically gauged by the level of bug-free performance that their applications provide. Talk about pressure! You have to face the people using your products the very next day after the product's release. In the Microsoft culture, this is called "eating your own dog food" and is an essential part of the company. The integration of TCP/IP into Microsoft involved the migration of 60,000 network nodes globally.

It was through these experiences that Microsoft realized that, while the first implementations of TCP/IP were robust enough for smaller work groups, they definitely needed work in terms of being able to scale accurately for the thousands of users both at Microsoft's headquarters and in regional offices throughout the world. After thorough internal testing of these protocols, both DHCP and WINS were integrated into Windows NT 4.0 and now Windows 2000. Microsoft is being very careful to not market a proprietary solution, so the RFCs for the DHCP and WINS implementations have been presented to the Internet Engineering Task Force (IETF).

With the latest release of TCP/IP in Windows 2000, Microsoft set the objectives of making their interpretation standards compliant and capable of being portable across platforms. Specifically, the design objectives as defined by Microsoft's design teams included making TCP/IP accomplish the following:

1. Standards compliant
2. Interoperable
3. Portable
4. Scalable
5. High performance
6. Versatile
7. Self-tuning
8. Easy to administer
9. Adaptable

The continual development of networking expertise at Microsoft has yielded further differentiation between its operating systems and specific command support within each operating system. Due to networking being so pervasive, Microsoft has recognized the need to include networking in each version of its operating system (Windows 95/98, Windows 2000, and Windows 2000 Server) at different levels of functionality.

GETTING TCP/IP UP AND RUNNING

Due to the extensive graphical interface of Windows 2000 Professional, installing TCP/IP support is relatively easy. You'll find that you have greater control over the configuration of options due to the use of numerous tabs in the Network dialog box. Having these tabs makes your approach easier when configuring network options.

As in Windows 95 and 98, Windows 2000 has three networking components: adapters, protocols, and services. These types form a stack corresponding to the overall network architecture; the services run on top, passing application data through the protocols, which code it for transmission via the adapter.

At a minimum, you'll need to have network adapters and the corresponding adapter driver installed in each Windows NT workstation that will be part of your network. In the case of integrating Novell NetWare into your network, you'll also want to have NWLink on each workstation. On each workstation you'll also want to have the NetBIOS interface and the RPC Name Service Provider services installed so that each workstation will be able to see other network-based systems. Many of the network services are optional and thus do not install automatically; these must be manually installed from the Control Panel. Some Windows NT services can be run with different protocols, while others are designed for use with a specific protocol.

HANDS-ON TUTORIAL: GETTING TCP/IP WORKING IN WINDOWS 2000 PROFESSIONAL

The real added value of having a network operating system is having the ability to integrate workstations of various operating systems together into a single, cohesive network. TCP/IP is the network protocol of choice for ensuring communication between workstation, servers, and, in short, any computer that needs to reside on a network and share resources. Throughout this section of the chapter we'll explore how to configure Windows 2000 Professional for use in a TCP/IP-based environment, including the fundamentals behind each of the functions included in each of the protocols defined.

INSTALLING TCP/IP

Due to Windows 2000 Professional using a browserlike shell for many of the properties and attributes that are configurable, the entire process of installing

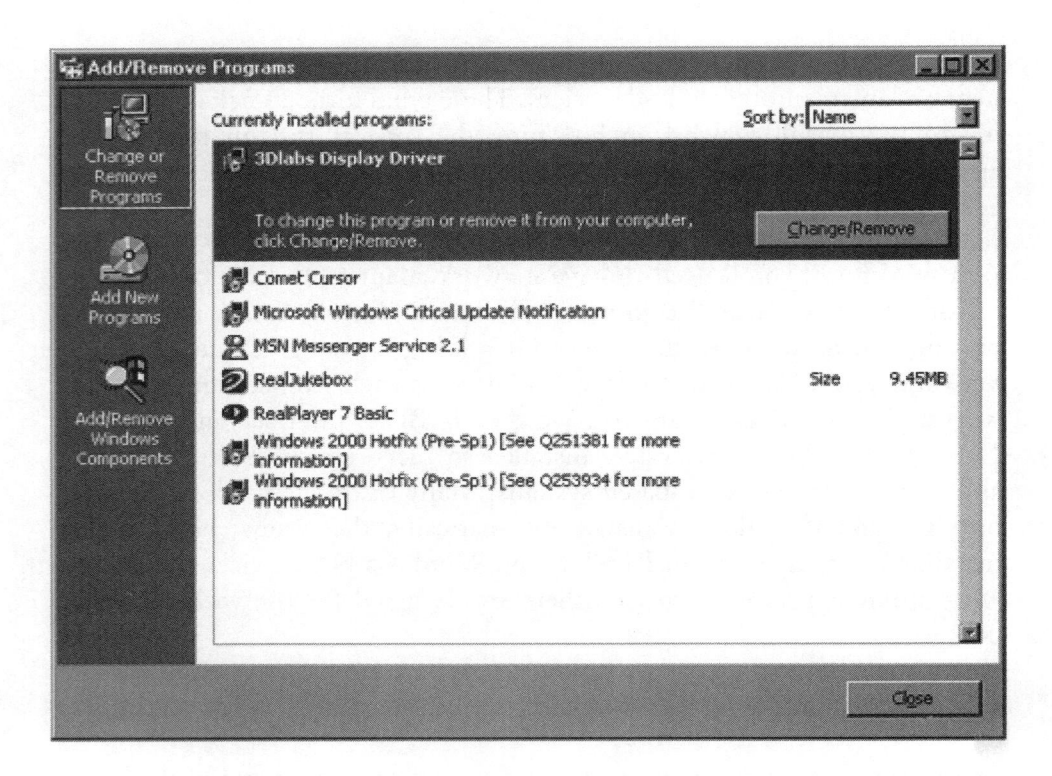

FIGURE 11.1 Loading Software from the Add/Remove Programs Wizard

and customizing TCP/IP is relatively simple. Where in Windows NT 4.0 the Network dialog box was found in the Control Panel, Windows 2000 Professional's networking is found in the Add/Remove Programs applet, also within the Control Panel. Once you select the Add Network Software entry in the Add/Remove Programs application, you then can configure a workstation for TCP/IP or for any other networking protocol supported in Windows NT. Figure 11.1 shows the Add/Remove Programs wizard, which is used when installing device drivers and software.

As mentioned earlier, the networking components included in Windows 2000 Professional are broken down into three types: adapters, protocols, and services. These types form a stack corresponding to the overall networking architecture itself. The services run on top, passing application data through the protocols, which code it for transmission via the adapter.

Configuring TCP/IP for DHCP Support

You'll find that Windows 2000 Professional has significantly different navigational options for getting to the DHCP options available within the context of a TCP/IP network session. DHCP is used for allocating Internet addresses dynamically as a system user dials in to use the network. It is considered to be one of the strongest differentiators the Windows NT operating system has—Microsoft's implementation is now three product generations old. DHCP relies on client/server architecture the same as the other components of the TCP/IP command set in Windows 2000 Professional and Windows 2000 Server. DHCP is considered to be one of the command options of the TCP/IP command set. It is not a separate client, service, or protocol within Windows 2000 Professional; it is integral to the implementation of TCP/IP and is configured using the tools available for customizing the TCP/IP definition.

Follow the steps below to get DHCP configured within Windows 2000 Professional.

1. Access the Network Connections window. Windows 2000 Professional has a new graphical appearance, and the dialog boxes you are accustomed to using are one level lower in its overall navigational structure. You can access the Network Connections window either from the Control Panel or by right-clicking on My Network Places on the desktop and selecting Properties. Either approach lands you in the Network Connection Wizard window, which is shown in Figure 11.2.

2. Click once on the button Assign TCP/IP addresses automatically using DHCP. Click once on OK after that.

You can also control DHCP services from the command line, using the following NET commands:

- NET START DHCPSERVER

- NET STOP DHCPSERVER

- NET PAUSE DHCPSERVER

- NET CONTINUE DHCPSERVER

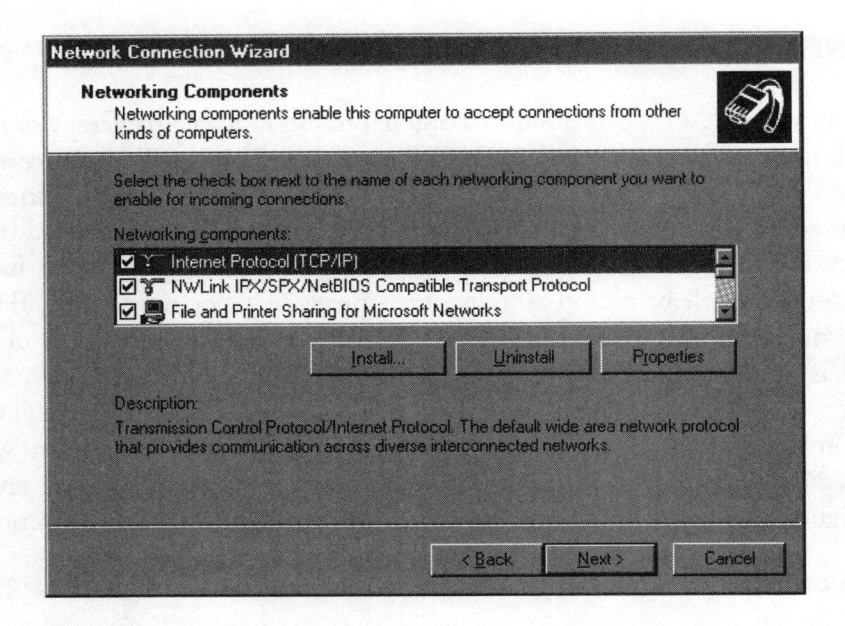

FIGURE 11.2 Using the Network Connection Wizard Window

2 Click once on the Properties... button. The next dialog box shown is the Incoming TCP/IP Properties dialog box (see Figure 11.3).

FIGURE 11.3 Local Area Connection Properties Dialog Box

Understanding Domain Name Services System

Created during the 1980s to handle the increasingly growing set of name resolution needs on the Internet, DNS was originally run on only a small set of computers. The network on which DNS was first tested was the now-famous ARPANET network of systems that linked many of the nation's campuses together. The earliest versions of DNS relied on a host table to resolve computer names into IP addresses. At the time, it was possible for every connected computer to have a complete listing of all computers on the internetwork. Changes were emailed to the Network Information Center at the Stanford Research Institute, which maintained the list and made it available to all users. Each user of the APRANET system had only to download the latest version every few weeks to stay current with all the names and associated IP addresses on the entire network.

As more and more computers were included on the ARPANET network of systems, the entries to DNS tables defining each member at their associated addresses grew quickly. ARPANET was the beginning of what would eventually turn into the Internet, with millions of users around the world; the volume of entries would quickly create a file that would take too much time to parse for address definition. Originally based on the logic of the first computers having an exhaustive list of every other system on the network, DNS has grown from the simple files used for arbitrating name resolution in ARPANET to a distributed name service and database system designed to spread the administration tasks around the network, dividing it into domains. Now, only domains are registered within the central repositories of the Internet, with InterNIC being the corporation that, in effect, rents unique domain addresses throughout the world. Unlike the early days of DNS where each individual system was defined within a DNS file, today only domain names are tracked centrally. The network administrator performs the assignment of a hostname to each workstation. As long as the administrator avoids creating duplicate names within the domain, the combination of the hostname (such as columlou) and the domain name (such as gateway.com) forms a fully qualified domain name (FQDN), which in this example is anony@gateway.com.

> **Quick Tip:** It's easy to get confused between the term domain used in conjunction with the Internet and a Windows NT domain. Although both

represent groups of computers, an Internet domain is registered with InterNIC to identify computers belonging to a particular company or organization. A Windows NT domain is a collection of computers located on a single internetwork that have been grouped for administrative efficiency.

Understanding Differences between DHCP and DNS

Both approaches to managing network IP addresses—DHCP and DNS—retain in their respective databases the unique identities of the systems that they are connecting to. There are a few differences between these protocols, however. We will profile them here:

- **Static versus Dynamic IP Addressing.** The first and most significant difference between DHCP and DNS is the method by which addresses are assigned. In networks configured with DHCP, the server assigns IP addresses by checking them out as a library checks out books out to members. In the same way, a DHCP-based network uses the assignment technique to ensure that each system gets a unique IP address. This approach to assigning IP addresses is called *dynamic IP addressing* because addresses are provided as each system on the network requires them.

    DNS-based networks are largely composed of systems that already have their IP addresses defined. The DNS entries within a DNS server reflect the systems on the network, like a phone book reflects persons living in a city. DNS is therefore a network arbitration approach relying on IP addresses being already in place.

- **System Identification during Network Operation.** Each system on a DHCP-based network uses an initialization sequence that alerts the DHCP server that it is ready to be assigned an IP address. The initialization sequence on the client systems of the network use a BOOT P protocol sequence to alert the DHCP server that an IP address needs to be defined. This BOOT P protocol was originally developed in the era of diskless workstations and continues today with TCP/IP-based network computers.

DNS uses an IP address that has been defined previously during a system's installation.

- **Content of Databases.** DHCP and DNS differ significantly in what is included in each of their respective databases. In terms of DCHP, the database acts as a central repository of IP addresses, which are then checked out on an as-needed basis. DHCP also enables clients to have a reserved, specific IP address in those instances when a company needs to mix both PCs that travel and those that stay stationary. DHCP in laptops provides the flexibility to get a connection at any time.

  In a DNS database, the names and addresses of each system are simply recorded because they have already been installed within the client systemroducing Simple Network Management Protocol.

If you're the system administrator for several dozen or even several hundred systems, you'll find that SNMP is a useful tool for managing your systems—it provides basic administrative information about devices attached to a network. SNMP is typically used by SNMP management software, which, in turn, takes advantage of the many desktop management initiatives (DMI) under way between Microsoft and Intel today. Specific to the needs of system administrators integrating Windows NT into their environments, SNMP is compatible with legacy DMI-based software tools. These include HP's OpenView, Netview, SNMPc, and several others.

What in essence does SNMP do? Its primary function is to enable communications between systems on a network and a host system. The host system—most often a server—has administrative tools installed that make it possible to monitor the overall health of the entire network by checking the performance and health of the client systems that comprise the network itself. SNMP is widely used and included in many peripherals and components of all types, including power conditioners, multiprocessor UNIX systems, and even motherboards themselves. One of the major developments in the SNMP arena is the integration of DMI-complaint firmware into the latest generation of dual-slot I and dual-slot II motherboards, which are Pentium II, Pentium II Xeon, and Pentium III compatible. Now you don't have to rely solely on the components to give you feedback on the overall performance of the system. Many of the motherboards available today, in conjunction with the functionality of Windows 2000, provide you with a baseline measure of system health so you can, in turn, as system administrator, manage the preventative maintenance on systems throughout your network. The integration of both SNMP in Windows 2000 and the DMI-complaint features in many motherboards today gives you a foundation on which to build. This enables you to create a comprehensive desktop management system where software components are loaded on top of SMNP to add definition and clarification to the systems being monitored from both the systems' components and the motherboard.

> **Quick Tip:** It's a good idea to get SNMP services running all the Windows NT–based systems you have because this service specifically provides object/counter relationships within Performance Monitor that make troubleshooting system performance that much easier. Why? With increased numbers of objects and associated counters, there is the potential of getting an increasingly insightful look into the performance of systems on your network.

INSTALLING SNMP

In this section we'll explore how to get SNMP up and running. These steps apply specifically to Windows 2000 Workstation and Server. For the Windows 2000 Professional configuration, use the Network icon in the Control Panel to access the network services needed for completing these steps.

1. Double-click on My Computer from the desktop. The My Computer window opens, showing all the storage devices on your system, in addition to icons for Network Connections, Scheduled Tasks, Control Panel, and Printers.

2. Double-click on the Control Panel icon, and the series of applets included are shown on screen. Notice that there is a shortcut icon for Network Connections within the Control Panel. If the system on which you are performing these steps hasn't had a network connection installed on it yet, the shortcut for Network Connections may not be present. In case this happens, go back to the active desktop (Windows 2000's main desktop) and right-click on the icon My Network Places. The Network Connections dialog box will appear.

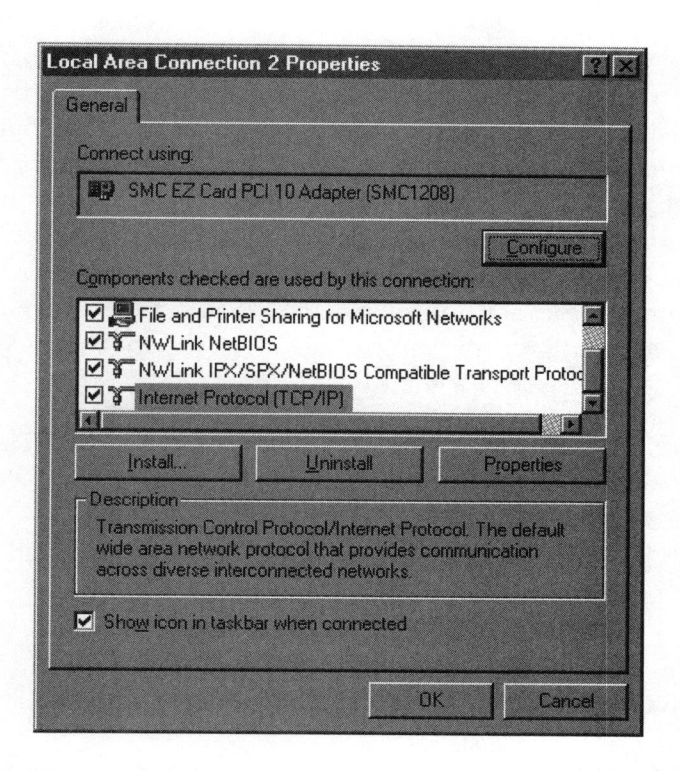

FIGURE 11.4 The Local Area Connection 2 Properties Dialog Box

For adding SNMP support for an existing network connection, right-click on the Network Connections icon.

3. Right-click on the Local Area Connection icon, selecting Properties from the shorthand menu that appears adjacent to the icon selected. The Local Area Connection 2 Properties dialog box appears (see Figure 11.4).

4. Click once on Install. The Select Network Component Type dialog box next appears, with three options to select from: Client, Service, and Protocol (see Figure 11.5).

5. Click once on Protocol, and then click once on Add. The Select Network Protocol dialog box appears.

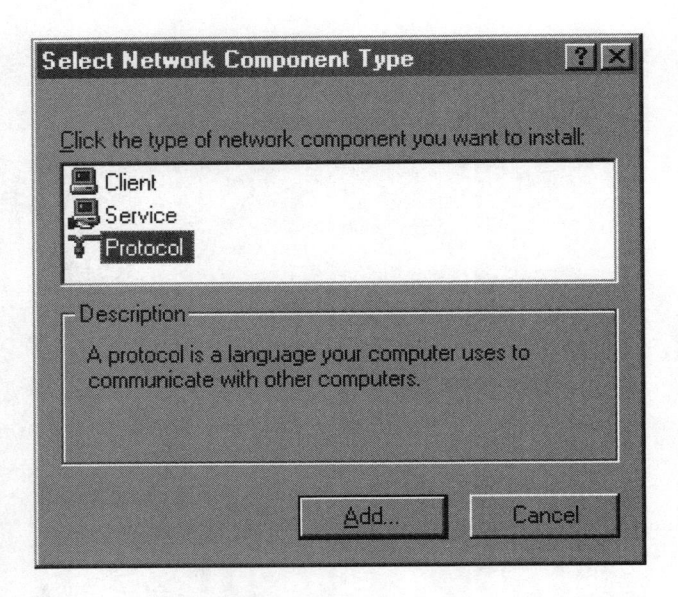

FIGURE 11.5 The Network Component Type Dialog Box (New to Windows 2000)

6. Select SNMP from the list of protocols; then click once on OK.

7. Click once on Cancel to close the Select Network Component Type dialog box.

8. Click once on OK to close the Local Area Connection Properties dialog box. The SNMP protocol has now been defined.

9. Close all applications and press Control+Alt+Delete.

10. Using the Windows NT Security dialog box, log off and then log on again to get the SNMP selections placed into the registry.

Once SNMP has been installed, you'll be able to see objects and counters specific to this protocol within the Performance Monitor's options. The Performance Monitor is included in Windows 2000 as well as previous versions of Windows NT.

ARCHITECTURE OF THE WINDOWS INTERNET NAMING SERVICES

Each workstation on a network has its own IP address and name when WINS is in effect in a network. When a network is configured using WINS, each new workstation on the network broadcasts its name to all other computers on the network, and an entry for that computer is inserted into the global name database for the network. This system works well on local networks where all protocols are supported by Microsoft network products. Any Microsoft operating system configured with TCP/IP protocols can also communicate with NetBIOS names connected by dissimiliar networks.

A significant limitation of the NetBIOS naming convention is that the names do not propagate across routers; NetBIOS names are disseminated using broadcast datagrams, which IP routers do not forward. Therefore, the NetBIOS names on one network are invisible to computers on other networks connected via routers.

Prior to the introduction of Windows NT Server 3.51, the Microsoft LAN Manager product supported internetwork name resolution using static naming tables stored in the LMHOSTS file. This was a direct result of the role of the etc/hosts file used so predominantly in UNIX-based networking routines found throughout sites using both LAN Manager and UNIX in mixed environments. An LMHOSTS file is a text file that contains mapping between NetBIOS names and IP addresses. In many instances where interoperability was needed between different systems, system administrators had to manually update LMHOSTS files by entering the IP addresses and system names of systems on the internetwork. This approach was a first-line-of-support type of activity to ensure connectivity between systems, especially between UNIX-based systems. With this labor-intensive approach to handling compatibility, it would sometimes take a system administrator hours to resolve connectivity issues.

With the introduction of Windows NT Server, Microsoft announced the Windows Internet Name Service. Like LMHOSTS, WINS maintains a NetBIOS global naming service for TCP/IP-based connections. Unlike LMHOSTS, WINS is dynamic, extending the automatic configuration of the NetBIOS name directory for local networks. The WINS database is updated automatically as NetBIOS systems insert and remove themselves from the network. If your Windows NT network will be connected to the Internet, it's possible to use WINS in conjunction with DNS, which enables WINS to provide DNS with hostnames for Microsoft-based hosts within your network.

The intent of the following section is to define the intricacies of WINS and the steps you can take to ensure smooth functioning of a network based on this protocol.

Resolving Names on Microsoft Networks

Within WINS, each of the names and their associated IP addresses need to be resolved. Providing the logic within the protocol for handling name resolution, NetBIOS handles the task of taking NetBIOS names in TCP/IP environments and referencing or resolving their relative identities on the network. NetBIOS over TCP/IP service (NBT) has evolved from a basic broadcast-based approach to the current name-service approach. Before looking into WINS in detail, it's important to get a handle on the name resolution modes supported by NBT. The b-node, p-node, and m-node resolution modes are defined in RFCs 1001 and 1002. H-node is being finalized.

- **B-node (broadcast name resolution mode).** Name resolution using broadcast messages is the oldest method that has been used for managing NBT name resolution on networks. B-name resolution works well in small, local networks, but has several disadvantages that become critical as networks grow. The first is that, as the number of hosts on the network increases, the increasing amount of broadcast traffic can consume significant network bandwidth. Second, IP routers do not forward broadcasts, and the b-node technique cannot propagate names throughout an internetwork. This node is the default name resolution mode for Microsoft hosts not configured for WINS name resolution. In pure b-node environments, hosts can be configured using LMHOSTS files to resolve names on remote networks.

- **P-node (directed name resolution mode).** Hosts configured for p-node use WINS for name resolution. P-node computers register themselves with an available WINS server, which functions as a NetBIOS name server. The WINS server maintains a database of the NetBIOS names, ensures that duplicate names do not exist, and makes the database available to WINS clients.

    Each WINS client is configured with the address of a WINS server, which may reside on the local network or on a remote network. WINS clients and servers communicate via directed messages than can be routed. No broadcast messages are required for p-node name resolution. Despite the strengths of p-node name resolution, it has two liabilities:

    — All computers using p-node must be configured using the address of a WINS server, even when communicating hosts reside on the same network.

    — If a WINS server is unavailable, name resolution fails for p-node clients.

Because both b-node and p-node address resolution present disadvantages, two address modes have been developed that form hybrids of b-node and p-node. These hybrid modes are called m-node and h-node.

- **M-node.** M-node computers first attempt to use b-node (broadcast) name resolution, which succeeds if the desired host resides on the local network. If b-node resolution fails, m-node hosts then attempt to use p-node to resolve the name.

    M-node enables name resolution to continue on the local network when WINS servers are down. B-node resolution is attempted first on the assumption that, in most environments, hosts communicate most often with hosts on their local networks. When this assumption holds, performance of the b-node resolution is superior to m-node. Recall, however, that b-node can result in a high level of broadcast traffic. Microsoft has also warned through their ATEC centers and the MCSE program that m-node can cause problems when network logons are attempted in a routed environment.

- **H-node.** Like m-node, h-node is a hybrid of broadcast (b-node) and directed (p-node) name resolution modes. Nodes configured with m-node, however, first attempt to resolve addresses using WINS. Only after an attempt to resolve the name using a name server fails does an h-node computer attempt to use b-node. M-node computers, therefore, can continue to resolve local addresses when WINS is unavailable. When operating in b-node mode, m-node computers continue to poll the WINS server and revert to h-node when WINS services are restored. H-node is the default mode for Microsoft TCP/IP clients configured using the addresses of WINS servers. As a fallback, Windows TCP/IP clients can be configured to use LMHOSTS files for name resolution.

Quick Tip: Although networks can be configured using mixtures of b-node and p-node computers, Microsoft recommends this as only an interim measure. P-node hosts ignore b-node broadcast messages, and b-node hosts ignore p-node directed messages. Two hosts, therefore, can conceivably be established using the same NetBIOS name.

NAMING ON A WINS NETWORK

Once a WINS server has been configured and the all workstations have been registered with the WINS database, attempts to connect with remote systems will be successful. WINS relies on the database to resolve naming and identity throughout the network. The function of the database is to respond back with the appropriate address from the WINS database to enable the connection between server and workstations.

MANAGING WINS SERVERS

WINS functions are managed using the WINS Server Manager. The icon for WINS Server Manager is installed into the Network Administration group of applets in Windows 2000 Server. WINS Server Manager is used for monitoring WINS servers, establishing static address mappings, and managing database replication. A few WINS database management tasks, such as compacting the database, are initiated from the command line. Like previous versions of Windows NT, the command lines are used for invoking the WINS Server Manager functions.

ADDING WINS SERVER TO WINS SERVER MANAGER

Let's look at one of the first tasks you'll use the WINS Server Manager to complete. WINS Server Manager can be configured for several servers at the same time. A single WINS Server Manager can also be used to monitor all WINS servers on the Internet.

If WINS Server Manager is run on a workstation running the WINS Server service, the workstation is listed in the WINS servers list. To add a WINS server to the last of managed servers, follow this series of steps:

1. Choose the Add WINS Server command from the Servers menu to open the Add WINS Server dialog box.

2. Enter the IP address of the new WINS server in the WINS Server entry box and choose OK. The server is added to those in the WINS Servers box.

To remove a WINS server from the list, select the server and choose the Delete WINS Server command in the Server menu. You can also use the Microsoft Management Console to accomplish this.

MONITORING WINS

The main window of the WINS Server Manager displays several statistics about the WINS server selected in the WINS servers box:

- **Database Initialized.** Static mappings can be imported from LMHOSTS files. This value indicates when static mappings were last imported.

- **Server Start Time.** The date and time when the WINS server is turned on and used for managing a network. This is the time the computer was started. Stopping and starting the WINS Server service does not reset this value.

- **Statistics Cleared.** The date and time when the server's statistics were cleared with the Clear Statistics command in the View menu.

- **Last Replication Time: Periodic.** The last time a WINS database replication was forced by an administrator.

- **Last Replication Time: Net Update.** The last time the WINS database was updated in response to a push request from another WINS server.

- **Total Queries Received.** The number of name queries this WINS server has received from WINS clients. Statistics indicate the number of names the WINS server successfully released and the number that it failed to release.

- **Total Releases.** The number of messages indicating an orderly shutdown of a NetBIOS application. Statistics indicate the number of names the WINS server released successfully and the number that it failed to release.

- **Total Registrations.** The number of registration messages received from clients.

When using the WINS Server Manager, you can also press the F5 key and have the entire array of variables refreshed to the most accurate values.

VIEWING WINS SERVER DETAILS

Detailed information for each WINS server may be displayed by selecting the server and choosing the Detailed Information command in the Server menu. The fields of the Install WINS Server dialog box are defined here.

- **Computer Name.** The NetBIOS name of the computer supporting the WINS server.

- **IP Address.** The IP address of the WINS server.

- **Connected Via.** The connection protocol.

- **Connected Since.** The time when the WINS Server service was last activated. Unlike the Server Start Time statistic in the main window, Connected Since is the reset when the WINS Server service is stopped or started.

- **Last Address Change.** The time when the last database change was replicated.

- **Last Scavenging Times.** The last time the database was scavenged to remove old data. Times are reported for the following scavenging events:

  — **Periodic.** Timed scavenging.

  — **Admin Trigger.** Manually initiated scavenging.

  — **Extinction.** Scavenging of released records that were scavenged because they had aged past their extinction time.

  — **Verification.** Last scavenging based on the Verify interval in the WINS server configuration.

- **Unique Registrations.** The number of name registrations for groups that the WINS server has accepted. The Conflicts statistic indicates the number of conflicts encountered when registering names that are already registered. The Renewals statistic indicates the number of renewals that have been received for unique names.

- **Group Registrations.** The number of requests for groups that the WINS server has accepted. The Conflicts statistic indicates the number of conflicts encountered when registering group names. The Renewals statistic indicates the number of group name renewals that have been received.

Configuring Static Mappings

Sometimes having dynamic name-address mappings are not desirable. This is particularly true when you have a group of users who are going to be in the same location for a long period of time. At such times, creating static mappings in the WINS database proves useful. A *static mapping* is a permanent mapping of a computer name to an IP address. Static mappings cannot be challenged and are removed only when they are explicitly deleted.

> **Note:** Reserved IP addresses assigned to DHCP clients override any static mappings assigned by WINS.

To add static mappings in WINS Server Manager, use the following steps:

1. Choose Static Mappings in the Mappings menu to open the Static Mappings dialog box that lists all active static mappings. The mappings for each specific system are tagged by a group icon defined using multihomed mapping.

2. To add a static mapping, choose Add Mappings to open the Add Static Mappings dialog box.

3. Type the workstation name in the Name box. WINS Server Manager supplies the \\ characters to complete the uniform naming convention (UNC) name.

4. Enter the network address in the IP Address box.

5. Choose one of the buttons in the Type box. The following choices are available:

   — **Unique.** The name will be unique in the WINS database and will have a single IP address.

   — **Group.** Groups are targets of broadcast messages and are not associated with IP address.

   — **Internet Group.** A group associated with the IP addresses of up to 24 Windows 2000 Server locations, plus the address of the primary domain controller, for a total of 25 Windows 2000 Server locations.

   — **Multihomed.** A name that can be associated with up to 25 addresses, corresponding to the IP addresses of a multihomed computer.

6. Choose Add.

You can also edit static mappings by performing the following steps:

1. Choose Static Mappings in the Mappings menu.

2. Select the mapping to be modified in the Static Mappings dialog box and choose Edit Mapping.

3. In the Edit Static Mapping dialog box, make required changes.

4. Choose OK to save the changes.

Static mappings for unique and special group names can be imported from files that conform to the format of the LMHOSTS files, described later in this chapter.

Managing the WINS Database

Replicating the WINS Database

Having two or more WINS servers on a network is very desirable, because you can use a second server to maintain a replica of the first WINS database; you can use the second server if the primary server fails. On large internetworks, multiple WINS servers result in less routed traffic and spread the name resolution workload across several computers.

Pairs of WINS servers can be configured as replication partners; WINS servers can perform two types of replication actions: pushing and pulling. A member of a replication pair functions as either a push partner or a pull partner.

All database replication takes place by transferring data from a push partner to a pull partner. A push partner cannot unilaterally push data. Data transfers may be initiated in two ways:

- A pull partner can initiate replication by requesting replication from a push partner. All records in a WINS database are stamped with a version number. When a pull partner sends a pull request, it specifies the highest version number that is associated with data received from the push partner. The push partner then sends any new data in its database that has a higher version than the one specified in the pull request.

- A push partner can initiate replication by notifying a pull partner that the push partner has data to send. The pull partner indicates its readiness to receive the data by sending a pull replication request that enables the push partner to push the data.

In summary, replication cannot take place until a pull partner indicates it is ready to receive data. A pull request indicates a readiness to receive data as well as the data the pull partner is prepared to receive. Therefore, the pull partners really control the replication process. All data is transferred from a push partner to a pull partner. Data are pushed only in response to pull requests.

Pulls generally are schedule events that occur at regular intervals. Pushes are generally triggered when the number of changes to be replicated exceeds a specific threshold. However, an administrator can manually trigger both pushes and pulls.

In general, replication is configured for two-way record transfer. Each member of the partnership is configured as a push partner and a pull partner, enabling both servers to pull updated data from each other.

MAINTAINING THE WINS DATABASE

Once a WINS database is up and running, it generally requires little maintenance. As with any database-oriented connectivity tool, it does need periodic maintenance. For continual increased performance of a WINS database, the size of the individual files need to be optimized. Additionally, when clients experience name resolution problems, you might need to view the contents of the WINS database to diagnose problems.

Backing Up the WINS Database

WINS performs a complete backup of its database every 24 hours. The file name and path are specified by registry parameters. On occasion, you might want to execute an unscheduled backup. The procedure, which must be performed on the computer running WINS Server services, is as follows:

1. Choose the Backup Database command in the Mappings menu to open the Select Backup Directory dialog box.

2. If desired, select a disk drive in the Drives box. The best location is another hard disk so that the database files remain available if the primary hard disk fails.

3. Specify the directory in which you want the backup files to be stored. WINS Server Manager will propose a default directory.

4. If desired, specify a new directory name to be created in the directory shown in Step 3. By default, a subdirectory named wins_bak is created to store the backup files.

5. To back up only those records that have changed since the last backup, check the Perform Incremental Backup box. The option is meaningful only if a full backup has not been previously performed.

6. Choose OK to make the backup.

It's a good idea to also back up the registry entries related to WINS. To back up the WINS registry entries, perform the following steps:

1. Run the RGEDT32.EXE program from the Run prompt on the File menu of Program Manager or File Manager. REGDT32 can also be run from the Command prompt.

2. Select HKEY_LOCAL_MACHINE window.

3. Select the key SYSTEM\CurrentControlSet\Services\WINS.

4. Choose the Save Key command in the Registry menu.

5. Specify a directory and file name in which to store the backup files.

6. Choose OK.

Restoring the WINS Database

If users cannot connect to a server running the WINS Server service, the WINS database probably has become corrupted. In this case, you might need to restore the database from a backup copy. You can do this using menu commands or manually. The procedure must be performed on the computer running the WINS Server service.

To restore the WINS database using menu commands:

1. Stop the WINS Server service using one of these methods:

    — Stop the Windows Internet Server service using the Services tool in the Control Panel.

    — Open a Command prompt and enter the command net stop wins.

2. Start the WINS Server Manager. Ignore any warning message to the effect that Windows Internet Naming Services is not running on the target machine, or that the target machine is not accessible.

3. Choose the Restore Local Database command in the Mappings menu.

4. In the Select Directory to Restore From dialog box, specify the directory from which to restore.

5. Choose OK to restore the database.

6. Start the WINS Server service using one of the following methods:

    — Start the Windows Internet Server Service using the Services tool in the Control Panel.

    — Open a Command prompt and enter the command net start wins.

Managing LMHOSTS Files

Although a complete name resolution system can be based on an LMHOSTS file type of architecture, static naming files can be very difficult to administer, particularly when they must be distributed to several hosts on the network. Nevertheless, LMHOSTS files may be necessary if WINS will not be run on a network or if having a backup is desirable in the case of a WINS service stopping unexpectedly or failing.

Although LAN Manager host files supported little more than mappings of NetBIOS names of IP addresses, Windows 2000 offers several options that make LMHOSTS considerably more versatile.

Format of LMHOSTS Files

A sample LMHOSTS file is installed in the directory

```
C:\WINNT\SYSTEM32\DRIVERS\ETC
```

This file is typically edited to show both IP addresses and system names that refer to systems throughout the network with which your workstations need to communicate. LMHOSTS can list IP addresses and names that apply to UNIX-based systems as well.

The basic format of an LMHOSTS entry is as follows:

```
ip address name
```

The IP address must begin in column one of the line. Here is an example of a basic LMHOSTS file:

```
129.160.27.35           PASADENA

                        DUARTE

                        HOLLYWOOD
```

LMHOSTS Keywords

You can also use keywords to make LMHOSTS files easier to navigate. Here is an example of an LMHOSTS file augmented using keywords:

```
LAGUNA                    #PRE

SANCLEMENTE               #PRE        #DOM:      WESTERN OFFICE

#BEGIN_ALTERNATE

#INCLUDE    \\LAGUNA\PUBLIC\LMHOSTS

#INCLUDE    \\SANCLEMENTE\PUBLIC\LMHOSTS

#END_ALTERNATE
```

The #PRE keyword specifies that the entry should be preloaded into the name chance. Ordinarily, LMHOSTS is consulted for name resolution only after WINS and b-node broadcasts have failed. Preloading the entry ensures that the mapping will be available at the start of the name resolution process.

The #DOM keyword associates an entry with a domain, which might be useful for determining how browsers and login servers behave on a routed TCP/IP network. #DOM entries can be preloaded in the cache by including the #PRE keyboard.

The #INCLUDE keyword makes it possible to load mappings from a remote file. One use for #INCLUDE is to support a master LMHOSTS file stored on logon servers and accessed by TCP/IP clients during startup. Entries in the remote LMHOSTS file are examined only when TCP/IP is started. Entries in the remote LMHOSTS file, therefore, must be tagged with the #PRE keyword to force them to be loaded into the cache.

If several copies of the included LMHOSTS file are available on different servers, you can force the computer to search several \\ locations until a file is successfully loaded. This is accomplished by bracketing #INCLUDE keywords between the keywords #BEGIN_ALTERNATVE and #END_ALTERNATE, as was done in the example file just presented. Any successful #INCLUDE causes the group to succeed.

Enabling Clients to Use LMHOSTS Files

Generally speaking, files are unnecessary on networks that have a properly functioning WINS name service. If an internetwork does not use WINS, LMHOSTS lookups should be enabled and LMHOSTS files should be configured to enable computers to find critical hosts.

Any TCP/IP client can be enabled to use LMHOSTS files through the use of the Enable LMHOSTS Lookup box in the Advanced Microsoft TCP/IP Configuration dialog box.

Guidelines for Establishing LMHOSTS Name Resolution

B-node computers not configured to use WINS name resolution can use LMHOSTs to resolve names on remote networks. If the majority of name queries are on the local network, preloading mappings in the LMHOSTS file is generally not necessary. Frequently accessed hosts on remote networks can be preloaded using the #PRE keyword.

#DOM keyboards should be used to enable non-WINS clients to locate domain controllers on remote networks. The LMHOSTS file for every workstation in the domain should include #DOM entries for all domain controllers that do not reside on the local network. This ensures that domain activities such as logon authentication continue to function.

To browse a domain other than the logon domain, LMHOSTS must include the name and IP address of the primary domain controller of the domain to be browsed. Include backup domain controllers in the case the primary fails or a backup domain controller is promoted to primary. LMHOSTS files on backup domain controllers should include mappings to the primary domain controller name and IP address, in addition to mappings to all other backup domain controllers. All domain controllers in trusted domains should be included in the local LMHOSTS files of each server being used throughout the network.

SUMMARY

Microsoft continues to base much of its primary naming system on NetBIOS names. Each workstation on the network has its own specific name, and all other computers on the local network know what that name is. When the network acquires a new workstation, the global name database updates automatically. Consequently, system maintenance is relatively effortless when it comes to main computers and keeping track of system names. This system works well on local networks. Microsoft operating systems can use NetBIOS names within the context of a local, non-routed network.

One of the unfortunate shortcomings of the NetBIOS naming system is that the names do not propagate across routers. NetBIOS names are disseminated using broadcast datagrams, and IP routers do not forward them. Computers on one network cannot read the NetBIOS names on another network when the networks are connected via a router. Microsoft does offer an alternative to the NetBIOS naming system that allows for recognition of names across networks (LMHOSTS), but the maintenance of this file is significant.

Windows Internet Naming Service, or WINS, was specifically developed to allow for dynamic updating of a database of IP names, as well as to provide the flexibility of having IP addresses and coresponding names read on a global level.

CHAPTER 12

EVALUATING WIRELESS NETWORKS FOR YOUR HOME OR OFFICE

Wireless technologies are already setting the technological pace of change in many networking arenas, and there's little doubt that they will transform home and small business networking within the next few years as well. There are two dominant standards in wireless networking: Home RF and IEEE 802.11b. As with any set of standards, each has its own strengths and weaknesses. Each one has to be looked at in terms of whether it is an appropriate fit for each specific application. What's so alluring about wireless networks in the home is that they make it possible to integrate cellular phones into a home network and use to the phone as a control device for systems throughout the home. You would be able to order tickets for sporting events online, control your security system, customize lighting and music preferences for guests in your home, and even order music and movie selections in a home theater. Integrating cellular phones into a home network brings about the possibility of an entirely new level of commerce, sometimes called mobile commerce, or m-commerce for short. Originally developed by e-Market Dynamics, m-commerce is expected to have one of the highest levels of revenue growth among all businesses during the first five years of the new millennium.

FORECASTS BY E-MARKET DYNAMICS

The following assumptions were used by e-Market Dynamics to develop their subscriber and revenue forecasts, discussed in the last two sections.

- In 1999, only Sprint PCS offered any kind of m-commerce solution, and it only did so for the month of December. It is e-Market Dynamics' belief that, despite the holidays, not many subscribers chose to make purchases over their wireless handsets.

- In 2000, many carriers began offering smart phones and e-wallet solutions, making it easier to see the information and to pay for the product or service over the phone. In 2000 most carriers introduced a wireless ecommerce application; by the end of 2000, a variety of applications evolved.

- In order to have access to m-commerce applications, subscribers must have access to the Internet with their devices and the access must be two way.

- In 1999, Sprint PCS began the m-commerce revolution by offering stock trades using Ameritrade on the Sprint PCS network for $8 per trade. Sprint also offered book purchasing from Amazon.com over the network in December.

- In 2000, there were still no micropayment solutions. Books, stocks, auction items, and gift certificates were the main purchases. The value of each transaction in 2000 was somewhat higher than in 1999, and there were more of them each month.

- In 2001, the commonly purchased items also include plane tickets and other big-ticket items, so the value of each transaction is increasing again. However, by the end of 2001, some micropayment solutions will be available. This will drive down the average cost per transaction, but it will also greatly increase the number of transactions per month.

- From 2002 to 2004, more micropayment applications will emerge, further decreasing the average value of each transaction but increasing the number of transactions per month.

Subscriber Forecast

To create the subscriber forecasts for m-commerce, e-Market Dynamics first had to create a forecast of wireless Internet users. A wireless Internet user is defined as anyone who accesses information on a mobile device across a wireless network that originates from either Intranets or broader, external Internets. The user must have some level of interactivity with the information (which means that one-way, push-only services are not included). The user must use the service at least once a month to qualify. The forecasters came up with a percentage of those wireless Internet subscribers who are and will be m-commerce subscribers based on assumptions of the kinds of services that are and will be available in the future. This percentage was then applied to the number of wireless Internet subscribers to create a forecast of m-commerce subscribers.

Table 12.1 shows the number of wireless m-commerce subscribers split by the type of access—circuit-switched, packet, short message service (SMS), two-way paging, and mobile data networks.

Wireless packet data and SMS will be the most common networks used for m-commerce transactions throughout the forecast period, with SMR following closely behind. Overall, the m-commerce subscriber base will grow from fewer than 1,000 in 1999 to more than 29 million in 2004.

Revenue Forecast

To create the transaction value forecast, e-Market Dynamics assumed an average number of transactions that each m-commerce subscriber would make and the average value of each transaction. Although m-commerce subscribers must access data on their handsets at least once a month, they need to make only one purchase a year using the mobile device to be considered an m-commerce subscriber; therefore, the minimum number of transactions per year would be one.

Table 12.2 shows the forecast of the average monthly cost per m-commerce transaction from 1999 to 2004. In 1999 there were only a few m-commerce options—Ameritrade and Amazon.com. Although trading stocks on a wireless phone cost $8 per trade, the monthly transaction value is lower because

TABLE 12.1 U.S. Mobile Commerce Subscribers by Access, 1999–2004 (000)

	1999	2000	2001	2002	2003	2004	Compound Annual Growth Rate (CAGR) (%) 2000–2004
Circuit Technology							
(CCT)	0	134	592	1,077	1,559	2,060	97.9
Packet	0	73	1,701	4,009	7,083	10,321	244.3
SMS	—	80	679	2,461	6,230	13,176	258.3
SMR	—	58	287	704	1,284	2,036	143.7
Two-way paging	—	41	147	445	861	1,287	137.3
Movile Device Networks							
(MDN)	—	13	35	87	147	202	100.1
Total	0	399	3,441	8,783	17,164	29,081	192.3

Source: "Wireless Networking and Forecast," e-Market Dynamics, 2000.

Key assumptions:

- In 1999, only Sprint PCS offered any kind of m-commerce solution, and it did so only for the month of December. It is e-Market Dynamics' belief that, despite the holidays, not many subscribers chose to make purchases over their wireless handsets.
- In 2000 many carriers started to offer smart phones and electronic wallet (ewallet) solutions, making it easier to see the information and to pay for the product or service over the phone.
- In 2000 most carriers introduced a wireless ecommerce application; by the end of 2000, a variety of applications evolved.
- In order to have access to m-commerce applications, subscribers must have Internet access via their devices, and it must be two-way access.

Messages in the data:

- In 2004, 20% of wireless subscribers will be conducting m-commerce transactions.
- There will be over 29 million mcommerce subscribers in 2004.

TABLE 12.2 U.S. Average Monthly Value of Mobile Commerce Transaction per Subscriber, 1999–2004 ($)

							Compound Annual Growth Rate (CAGR) (%)
	1999	2000	2001	2002	2003	2004	2000–2004
Average value	5.20	12.00	25.00	40.00	60.00	75.00	58.1
Growth (%)	NA	130.8	108.3	60.0	50.0	25.0	—

Source: "Wireless Networking and Forecast," e-Market Dynamics, 2000.

Key assumptions:

- In 1999 most applications were stock trading and book purchases.
- n 2000 there were not yet any micropayment solutions. Applications included mainly books, stocks, auction items, and gift certificates.
- In 2001 and beyond, plane tickets and other big-ticket items will enter the picture, and various micropayment solutions will become available.

Messages in the data:

- The average value of each m-commerce transaction was quite low in 1999 because there were so few transactions per month. As the number of transactions increases, so will the average monthly spending.
- By 2004 each user will spend an average of $75 per month on m-commerce transactions.

it includes an average fee for the entire year of 1999, although these m-commerce applications were not even available until the fourth quarter. As shown in the table, the monthly value increases throughout the forecast period. Although e-Market Dynamics assumes micropayment applications will come about later in the forecast period, subscribers will be making a larger number of transactions each month, thereby increasing the monthly value of the transactions despite lower average values per transaction.

Table 12.3 presents the forecast of the annual value of m-commerce transactions from 1999 to 2004. Because applications did not become available until late 1999, and because few subscribers actually completed purchases

TABLE 12.3 U.S. Mobile Commerce Transaction Value, 1999–2004 ($M)

							Compound Annual Growth Rate (CAGR) (%)
	1999	2000	2001	2002	2003	2004	2000–2004
CCT	0	10	109	400	949	1,629	260.3
Packet	0	5	266	1,370	3,993	7,832	520.2
SMS	-	6	114	753	3,129	8,733	524.1
SMR	-	4	52	238	716	1,494	335.4
Two-way paging	-	3	28	142	470	967	326.5
MDN	-	1	7	29	84	157	262.8
Total	0	29	576	2,934	9,341	20,810	418.9

Source: "Wireless Networking and Forecast," e-Market Dynamics, 2000.

Key assumptions:

- The average monthly spending was low in 1999 because there were so few transactions per month. As the number of transactions increases, so will the average monthly spending.
- In order to have access to m-commerce applications, subscribers must have Internet access via their devices, and the access must be two way.

Message in the data:

- m-commerce will generate $20 billion in transactions in 2004.

when applications were first introduced, the value is so small that it does not show in the table. By 2004, m-commerce transactions will reach approximately $20 billion in the United States.

Because this is a forecast of the value of m-commerce transactions, obviously the carrier is not going to see the full transaction value as revenue. Instead,

TABLE 12.4 Potential U.S. Mobile Commerce Affiliate Fees ($M)

		1999	2000	2001	2002	2003
3% affiliate fees	0	1	17	88	280	624
5% affiliate fees	0	1	29	147	467	1,041

Source: "Wireless Networking and Forecast," e-Market Dynamics, 2000.

Key assumption:

- In most cases, carriers will be paid an affiliate fee between 3% and 5% of the transaction value.

Message in the data:

- Carriers can generate an additional $1 billion in 2004 from affiliate fees alone.

carriers will probably receive between 3% and 5% as an affiliate fee. Table 12.4 presents the dollar amount carriers can expect to see if they receive 3% or 5% of the transaction value. In 2000, carriers could expect only between $1.0 and $1.4 million; by 2004 wireless carriers should be earning $624 million to $1.0 billion over service revenues.

Major Forecast Assumptions

The following assumptions were used by e-Market Dynamics to develop their subscriber and revenue forecasts, discussed in the last two sections.

- In 1999, only Sprint PCS offered any kind of m-commerce solution, and it only did so for the month of December. It is e-Market Dynamics' belief that, despite the holidays, not many subscribers chose to make purchases over their wireless handsets.
- In 2000, many carriers began offering smart phones and ewallet solutions, making it easier to see the information and to pay for the product or service over the phone. In 2000 most carriers introduced

a wireless ecommerce application; by the end of 2000, a variety of applications evolved.

- In order to have access to m-commerce applications, subscribers must have access to the Internet with their devices and the access must be two way.

- In 1999, Sprint PCS began the m-commerce revolution by offering stock trades using Ameritrade on the Sprint PCS network for $8 per trade. Sprint also offered book purchasing from Amazon.com over the network in December.

- In 2000 there were still no micropayment solutions. Books, stocks, auction items, and gift certificateswere the main purchases. The value of each transaction in 2000 was somewhat higher than in 1999, and there were more of them each month.

- In 2001, the commonly purchased items also include plane tickets and other big-ticket items, so the value of each transaction is increasing again. However, by the end of 2001, some micropayment solutions will be available. This will drive down the average cost per transaction, but it will also greatly increase the number of transactions per month.

- From 2002 to 2004, more micropayment applications will emerge, further decreasing the average value of each transaction but increasing the number of transactions per month.

THE COMPETITIVE ADVANTAGE OF GOING WIRELESS

Today's business environment is characterized by an increasingly mobile workforce and flatter organizations. Employees are equipped with notebook computers and spend more of their time working in teams that cross functional, organizational, and geographic boundaries. Much of these workers' productivity occurs in meetings and away from their desks. Users need access to the network far beyond their personal desktops. WLANs fit well in this work environment, giving mobile workers much-needed freedom in their network access. With a wireless network, workers can access information from anywhere in the corporation—a conference room, the cafeteria, or a remote branch office. WLANs provide a benefit for information technology (IT) managers as well, enabling them to design, deploy, and enhance networks without regard to the availability of wiring, saving both effort and dollars.

Businesses of all sizes can benefit from deploying a WLAN system, which provides a powerful combination of wired network throughput, mobile access, and configuration flexibility. The economic benefits can add up to as much as $16,000 per user—measured in worker productivity, organizational efficiency, revenue gain, and cost savings—over wired alternatives.[1] Specifically, WLAN advantages include the following:

- Mobility that improves productivity with real-time access to information, regardless of worker location, for faster and more efficient decision making

- Cost-effective network setup for hard-to-wire locations such as older buildings and solid-wall structures

- Reduced cost of ownership—particularly in dynamic environments requiring frequent modifications—thanks to minimal wiring and installation costs per device and user

WLANs liberate users from dependence on hard-wired access to the network backbone, giving them anytime, anywhere network access. This freedom to roam offers numerous user benefits for a variety of work environments:

- Immediate bedside access to patient information for doctors and hospital staff

- Easy, real-time network access for on-site consultants or auditors

- Improved database access for roving supervisors such as production line managers, warehouse auditors, or construction engineers

---

[1] "Wireless Local Area Networking: ROI/Cost-Benefit Study," WLANA, October 1998.

- Simplified network configuration with minimal management information systems (MIS) involvement for temporary setups such as trade shows or conference rooms

- Faster access to customer information for service vendors and retailers, resulting in better service and improved customer satisfaction

- Location-independent access for network administrators, for easier on-site troubleshooting and support

- Real-time access to study group meetings and research links for students

IEEE 802.11 AND 802.11B TECHNOLOGY

As the globally recognized LAN authority, the IEEE 802 committee has established the standards that have driven the LAN industry for the past two decades, including 802.3 Ethernet, 802.5 Token Ring, and 802.3z 100Base-TX Fast Ethernet. In 1997, after seven years of work, the IEEE published 802.11, the first internationally sanctioned standard for WLANs. In September 1999 the committee ratified the 802.11 HR amendment to the standard, which added two higher speeds (5.5 and 11 Mbps) to 802.11. With 802.11b WLANs, mobile users can get Ethernet levels of performance, throughput, and availability. The standards-based technology enables administrators to build networks that seamlessly combine more than one LAN technology to best fit their business and user needs.

Like all IEEE 802 standards, the 802.11 standards focus on the bottom two levels of the OSI Model—the physical layer and data-link layer (see Figure 12.1). Any LAN application, network operating system, or protocol, including TCP/IP and Novell NetWare, will run on an 802.11-compliant WLAN as easily as it runs over Ethernet. The basic architecture, features, and services of 802.11b are defined by the original 802.11 standard. The 802.11b specification affects only the physical layer, adding higher data rates and more robust connectivity.

Application Services	**Layer 7 (Application)** - Communications-related services oriented towards specific applications. Examples include file transfer and email.	
	Layer 6 (Presentation) - Negotiates formats, transforms information into agreed-upon format, generates session requests for service	
	Layer 5 (Session) - Manages connections between cooperating applications by establishing and releasing sessions, synchronizing information transfer over these sessions, mapping session-to-transport and session-to-application sessions.	
Networking	**Layer 4 (Transport)** - Manages connections between two end nodes by establishing and releasing end-to-end connections; controlling the size, sequence, and flow of transport packets; mapping transport and network addresses.	
	Layer 3 (Network) - Routes information among source, intermediate, and destination nodes; establishes and maintains connections, if using connection-oriented exchanges or protocols.	
	Layer 2 (Data Link) - Transfers data frames over the physical layer; responsible for reliability.	
Transmission	**Layer 1 (Physical)** - Mechanical, electrical, functional, and procedural aspects of data circuits among network nodes.	

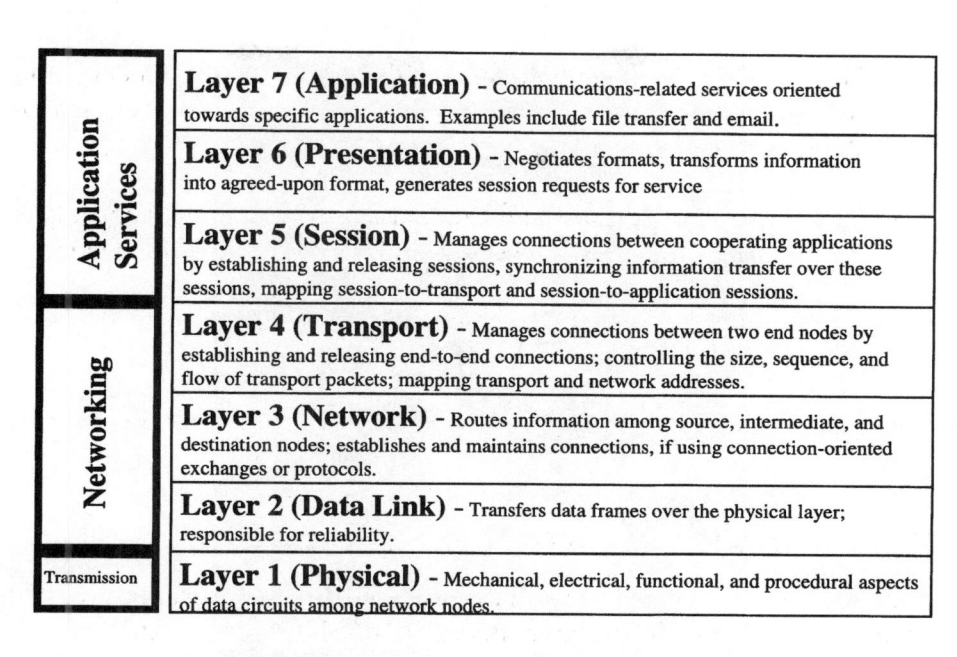

FIGURE 12.1 802.11 and the OSI Model

802.11 Operating Modes

802.11 defines two pieces of equipment—a wireless *station*, which is usually a PC equipped with a wireless NIC, and an *access point*, which acts as a bridge between the wireless and wired networks. An access point usually consists of a radio, a wired network interface (e.g., 802.3), and bridging software conforming to the 802.1d bridging standard. The access point acts as the base station for the wireless network, aggregating access for multiple wireless stations onto the wired network. Wireless end stations can be an 802.11 PC card, PCI, or ISA NICs, or they can be embedded solutions in non-PC clients (such as an 802.11-based telephone handset).

The 802.11 standard defines two modes: *infrastructure* mode and *ad hoc* mode. In infrastructure mode (see Figure 12.2), the wireless network consists of at least one access point connected to the wired network infrastructure and a set of wireless end stations. This configuration is called a Basic Service Set (BSS). An Extended Service Set (ESS) is a set of two or more BSSs forming

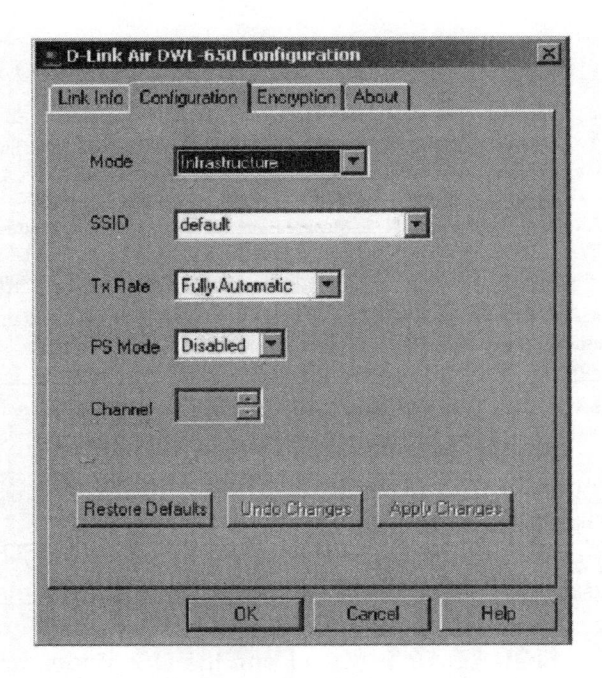

FIGURE 12.2 Infrastructure Mode

a single subnetwork. Because most corporate WLANs require access to the wired LAN for services (file servers, printers, Internet links), they will operate in infrastructure mode.

Ad hoc mode (also called peer-to-peer mode or an Independent Basic Service Set [IBSS]) is simply a set of 802.11 wireless stations that communicate directly with one another without using an access point or any connection to a wired network (see Figure 12.3). This mode is useful for quickly and easily setting up a wireless network anywhere that a wireless infrastructure does not exist or is not required for services, such as a hotel room, convention center, or airport, or where access to the wired network is barred (such as for consultants at a client site).

THE 802.11 PHYSICAL LAYER

The three physical layers originally defined in 802.11 include two spread-spectrum radio techniques and a diffuse IR specification. The radio-based

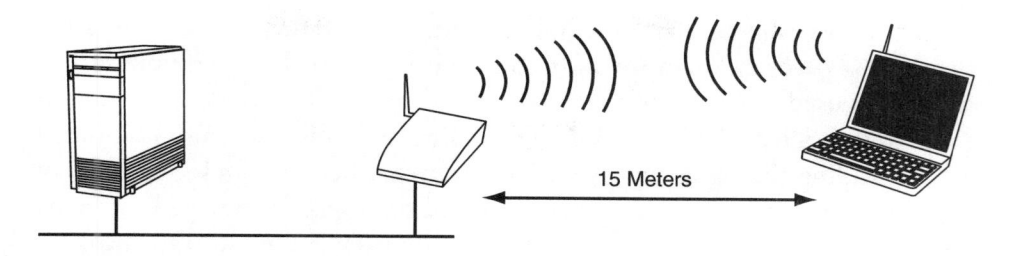

FIGURE 12.3 Ad Hoc Mode

standards operate within the 2.4-GHz ISM band. These frequency bands are recognized by international regulatory agencies, such as the FCC (United States), European Telecommunications Standards Institute (ETSI) (Europe), and the Japanese telecommunications standards committee, MKK. for unlicensed radio operations. As such, 802.11-based products do not require user licensing or special training. Spread-spectrum techniques, in addition to satisfying regulatory requirements, increase reliability, boost throughput, and allow many unrelated products to share the spectrum without explicit cooperation and with minimal interference. The original 802.11 wireless standard defines data rates of 1 Mbps and 2 Mbps via radio waves using FHSS or DSSS. It is important to note that FHSS and DSSS are fundamentally different signaling mechanisms and will not interoperate with one another.

Using the frequency-hopping technique, the 2.4-GHz band is divided into 75 subchannels of 1 MHz each. The sender and receiver agree on a hopping pattern, and data is sent over a sequence of the subchannels. Each conversation within the 802.11 network occurs over a different hopping pattern, and the patterns are designed to minimize the chance of two senders using the same subchannel simultaneously. FHSS techniques allow for a relatively simple radio design, but are limited to speeds of no higher than 2 Mbps. This limitation is driven primarily by FCC regulations that restrict subchannel bandwidth to 1 MHz. These regulations force FHSS systems to spread their usage across the entire 2.4-GHz band, meaning they must hop often, which leads to high hopping overhead.

In contrast, the direct-sequence signaling technique divides the 2.4-GHz band into 14 channels of 22 MHz each. Adjacent channels overlap one another partially, with 3 of the 14 being completely nonoverlapping. Data is sent across one of these 22-MHz channels without hopping to other channels. To compensate for noise on a given channel, a technique called *chipping* is used. Each bit of user data is converted into a series of redundant bit patterns called *chips*. The inherent redundancy of each chip, combined with spreading the signal across the 22-MHz channel, provides for a form of error checking and correction; even if part of the signal is damaged, in many cases it can still be recovered, minimizing the need for retransmissions.

802.11B Enhancements to the Physical Address (PHY) Layer

The key contribution of the 802.11b addition to the WLAN standard has been to standardize the physical layer support of two new speeds—5.5 Mbps and 11 Mbps. To accomplish this, DSSS had to be selected as the sole physical layer technique for the standard since, as noted above, frequency hopping cannot support the higher speeds without violating current FCC regulations. The implication is that 802.11b systems will interoperate with 1-Mbps and 2-Mbps 802.11 DSSS systems, but will not work with 1-Mbps and 2-Mbps 802.11 FHSS systems.

The original 802.11 DSSS standard specifies 11-bit chipping—called a *Barker sequence*—to encode all data sent over the air. Each 11-chip sequence represents a single data bit (1 or 0) and is converted to a waveform, called a *symbol*, that can be sent over the air. These symbols are transmitted at a *symbol rate* of 1 million symbols per second (1 s) using a technique called Binary Phase Shift Keying (BPSK). In the case of 2 MSps, a more sophisticated implementation called Quadrature Phase Shift Keying (QPSK) is used; it doubles the data rate available in BPSK via improved efficiency in the use of the radio bandwidth.

To increase the data rate in the 802.11b standard, advanced coding techniques are employed. Rather than the 2 11-bit Barker sequences, 802.11b specifies Complementary Code Keying (CCK), which consists of a set of 64 8-bit code words. As a set, these code words have unique mathematical properties that allow them to be correctly distinguished from one another by a receiver even in the presence of substantial noise and multipath interference (e.g., interference caused by receiving multiple radio reflections within a building). The 5.5-Mbps rate uses CCK to encode 4 bits per carrier, while

Table 12.5 802.11b Data Rate Specifications

Data Rate	Code Length	Modulation	Symbol Rate	Bits/Symbol
1 Mbps	11 (Barker Sequence)	BPSK	1 MSps	1
2 Mbps	11 (Barker Sequence)	QPSK	1 MSps	2
5.5 Mbps	8 (CCK)	QPSK	1.375 MSps	4
11 Mbps	8 (CCK)	QPSK	1.375 MSps	8

the 11-Mbps rate encodes 8 bits per carrier. Both speeds use QPSK as the modulation technique and signal at 1.375 MSps. This is how the higher data rates are obtained. Table 12.5 shows the differences.

To support very noisy environments as well as extended range, 802.11b WLANs use *dynamic rate shifting*, enabling data rates to be automatically adjusted to compensate for the changing nature of the radio channel. Ideally, users connect at the full 11-Mbps rate. However, when devices move beyond the optimal range for 11-Mbps operation, or if substantial interference is present, 802.11b devices will transmit at lower speeds, falling back to 5.5, 2, and 1 Mbps. Likewise, if the device moves back within the range of a higher-speed transmission, the connection will automatically speed up again. Rate shifting is a physical-layer mechanism transparent to the user and the upper layers of the protocol stack.

The 802.11 Data-Link Layer

The 802.11 standard data-link layer consists of two sublayers: LLC and MAC. 802.11 uses the same 802.2 LLC and 48-bit addressing as other 802 LANs, allowing for very simple bridging from wireless to IEEE wired networks, but the MAC is unique to WLANs.

The 802.11 MAC is very similar in concept to 802.3, in that it is designed to support multiple users on a shared medium by having the sender sense the medium before accessing it. For 802.3 Ethernet LANs, the Carrier Sense Multiple Access with Collision Detection (CSMA/CD) protocol regulates how Ethernet stations establish access to the wire and how they detect and handle collisions that occur when two or more devices try to communicate over the LAN simultaneously. In an 802.11 WLAN, collision detection is not possible due to what is known as the *near/far* problem: To detect a collision, a station must be able to transmit and listen at the same time, but in radio systems the transmission drowns out the ability of the station to "hear" a collision. To account for this difference, 802.11 uses a slightly modified protocol known as Carrier Sense Multiple Access with Collision Avoidance (CSMA/CA) or the Distributed Coordination Function (DCF). CSMA/CA attempts to avoid collisions by using explicit packet acknowledgment (ACK), which means an ACK packet is sent by the receiving station to confirm that the data packet arrived intact.

CSMA/CA works as follows: A station wishing to transmit senses the air, and, if no activity is detected, the station waits an additional, randomly selected period of time and then transmits if the medium is still free. If the packet is received intact, the receiving station issues an ACK frame that, once successfully received by the sender, completes the process. If the ACK frame is not detected by the sending station, either because the original data packet was not received intact or because the ACK was not received intact, a collision is assumed to have occurred, and the data packet is transmitted again after waiting another random amount of time. CSMA/CA thus provides a way of sharing access over the air. This explicit ACK mechanism also handles interference and other radio-related problems very effectively. However, it does add some overhead to 802.11 that 802.3 does not have, so that an 802.11 LAN will always have slower performance than an equivalent Ethernet LAN.

Another MAC-layer problem specific to wireless is the *hidden node* issue, in which two stations on opposite sides of an access point can each "hear" activity from an access point, but cannot hear activity from each other, usually due to distance or an obstruction. To solve this problem, 802.11 specifies an optional Request to Send/Clear to Send (RTS/CTS) protocol at the MAC layer. When this feature is in use, a sending station transmits an RTS and waits for the access point to reply with a CTS. Because all stations in the network can hear the access point, the CTS causes them to delay any intended transmissions, enabling the sending station to transmit and receive an ACK without any chance of collision. Because RTS/CTS adds additional overhead to the network by temporarily reserving the medium, it is typically used only on the largest-sized packets, for which retransmission would be expensive from a bandwidth standpoint.

Finally, the 802.11 MAC layer provides for two other robustness features: CRC checksum and packet fragmentation. Each packet has a CRC checksum calculated and attached to ensure that the data was not corrupted in transit. This is different from Ethernet, where higher-level protocols such as TCP handle error checking. Packet fragmentation enables large packets to be broken into smaller units when sent over the air, which is useful in very congested environments or when interference is a factor, as larger packets have a better chance of being corrupted. This technique reduces the need for retransmission in many cases and thus improves overall wireless network performance. The MAC layer is responsible for reassembling fragments received, rendering the process transparent to higher-level protocols.

ASSOCIATION, CELLULAR ARCHITECTURES, AND ROAMING. The 802.11 MAC layer is responsible for how a client associates with an access point. When an 802.11 client enters the range of one or more access points, it chooses an access point to associate with (also called joining a Basic Service Set) based on signal strength and observed packet error rates. Once accepted by the access point, the client tunes to the radio channel to which the access point is set. Periodically it surveys all 802.11 channels in order to assess whether a different access point would provide it with better performance characteristics. If it determines that this is the case, it reassociates with the new access point, tuning to the radio channel to which the access point is set (see Figure 12.4).

Reassociation usually occurs because the wireless station has physically moved away from the original access point, causing the signal to weaken. In other cases, reassociation occurs because there is a change in radio characteristics in the building or due to high network traffic on the original access point. A function known as *load balancing* handles the latter problem—its primary job is to distribute the total WLAN load most efficiently across the available wireless infrastructure.

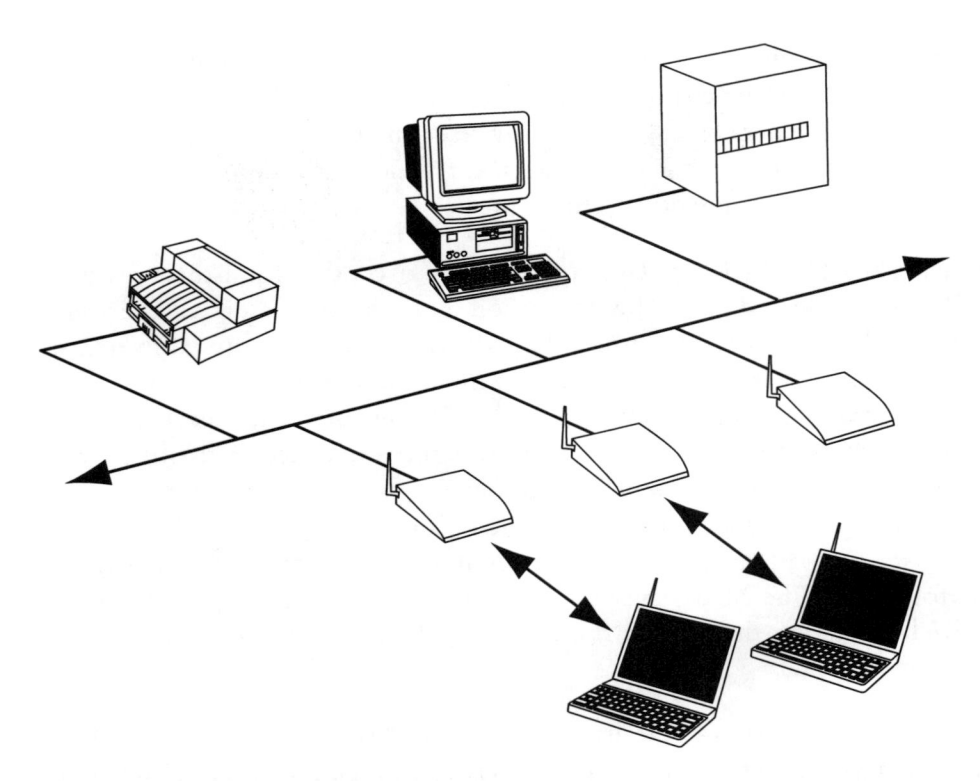

FIGURE 12.4 Access Point Roaming

This process of dynamically associating and reassociating with access points enables network managers to set up WLANs with very broad coverage by creating a series of overlapping 802.11b cells throughout a building or across a campus. To be successful, the IT manager ideally will employ *channel reuse*, taking care to set up each access point on an 802.11 DSSS channel that does not overlap with a channel used by a neighboring access point (illustrated in Figure 12.5). As noted above, while 14 partially overlapping channels are specified in 802.11 DSSS, only 3 channels do not overlap at all, and these are the best channels to use for multicell coverage. If two access points are in range of one another and are set to the same or partially overlapping channels, they may cause some interference for one another, thus lowering the total available bandwidth in the area of overlap.

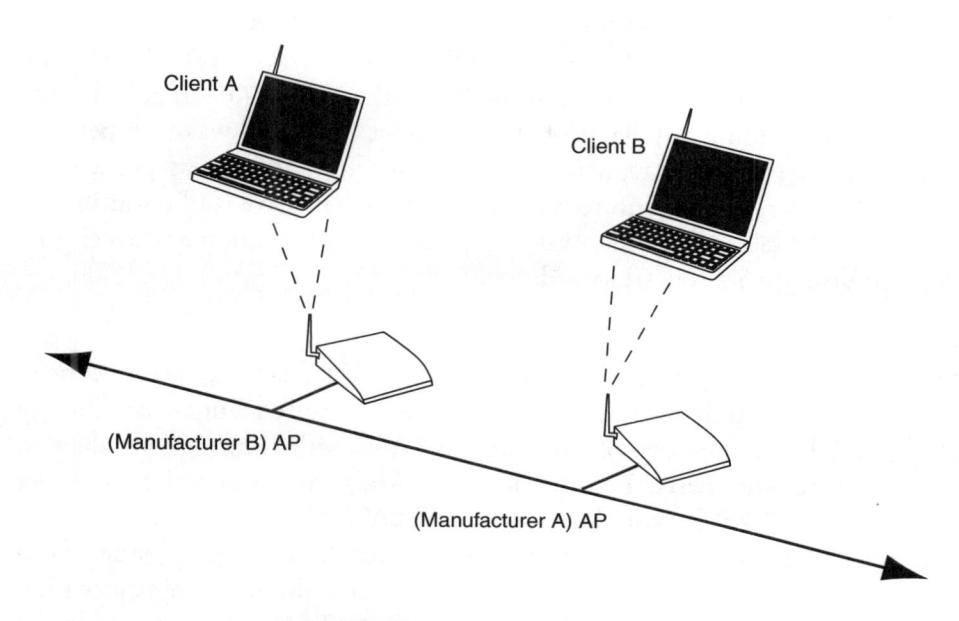

FIGURE 12.5 Unlimited Roaming

SUPPORT FOR TIME-BOUNDED DATA. Time-bounded data, such as voice and video, is supported in the 802.11 MAC specification through the Point Coordination Function (PCF). As opposed to the DCF where control is distributed to all stations, in PCF mode a single access point controls access to the media. If a BSS is set up with PCF enabled, the system spends half the time in PCF mode and half in DCF mode. During the periods when the system is in PCF mode, the access point will poll each station for data, and, after a given time, will move on to the next station. No station is allowed to transmit unless it is polled, and stations receive data from the access point only when they are polled. Because PCF gives every station a turn to transmit in a predetermined fashion, a maximum latency is guaranteed. A downside to PCF is that it is not particularly scalable, in that a single point needs to have control of media access and must poll all stations; this can be ineffective in large networks.

POWER MANAGEMENT. In addition to controlling media access, the 802.11 HR MAC supports power conservation to extend the battery life of portable devices. The standard supports two power-utilization modes, called Continuous Aware Mode and Power Save Polling Mode. In the former, the radio is always on and drawing power; in the latter, the radio is "dozing," with the access point queuing any data for it. The client radio will wake up periodically in time to receive regular *beacon* signals from the access point. The beacon includes information regarding which stations have traffic waiting for them, and the client can thus awake upon beacon notification and receive its data, returning to sleep afterward.

SECURITY. 802.11 provides for both MAC layer (OSI layer 2) access control and encryption mechanisms, which are known as Wired Equivalent Privacy (WEP), with the objective of providing WLANs with security equivalent to their wired counterparts. For the access control, the Extended Service Set Identification (ESSID, also known as a WLAN Service Area ID) is programmed into each access point and is required knowledge in order for a wireless client to associate with an access point. In addition, there is provision for a table of MAC addresses called an *Access Control List* to be included in the access point, restricting access to clients whose MAC addresses are on the list.

For data encryption, the standard provides for optional encryption using a 40-bit shared-key RC4 pseudorandom number generator (PRNG) algorithm from RSA Data Security. All data sent and received while the end station and access point are associated can be encrypted using this key. In addition, when encryption is in use, the access point will issue an encrypted challenge packet to any client attempting to associate with it. The client must use its key to encrypt the correct response in order to authenticate itself and gain network access.

Beyond layer 2, 802.11 HR WLANs support the same security standards that are supported by other 802 LANs for access control (such as network operating system logins) and encryption (such as IPSec or application-level encryption). These higher-layer technologies can be used to create end-to-end secure networks encompassing both wired LAN and WLAN components, with the wireless piece of the network gaining unique additional security from the 802.11 feature set.

CURRENT STANDARDS IN WIRELESS NETWORKING

What's New in Wireless LANs: The IEEE 802.11b Standard

A wireless LAN (WLAN) is a data transmission system designed to provide location-independent network access between computing devices by using radio waves rather than a cable infrastructure. In the corporate enterprise, WLANs are usually implemented as the final link between the existing wired network and a group of client computers, giving these users wireless access to the full resources and services of the corporate network across a building or campus setting.

WLANs are on the verge of becoming a mainstream connectivity solution for a broad range of business customers. The wireless market is expanding rapidly as businesses discover the productivity benefits of going wire-free. According to Frost and Sullivan, the WLAN industry exceeded $300 million in 1998 and will grow to $1.6 billion in 2005. To date, WLANs have been implemented primarily in vertical applications such as manufacturing facilities, warehouses, and retail stores. The majority of future WLAN growth is expected in healthcare facilities, educational institutions, and corporate enterprise office spaces. In the corporation, conference rooms, public areas, and branch offices are likely venues for WLANs.

The widespread acceptance of WLANs depends on industry standardization to ensure product compatibility and reliability among the various manufacturers. The IEEE ratified the original 802.11 specification in 1997 as the standard for WLANs. That version of 802.11 provides for 1-Mbps and 2-Mbps data rates and a set of fundamental signaling methods and other services.

The most critical issue affecting WLAN demand has been limited throughput. The data rates supported by the original 802.11 standard are too slow to support most general business requirements and this has slowed the adoption of WLANs. Recognizing the critical need to support higher data transmission rates, the IEEE recently ratified the 802.11b standard (also known as 802.11 High Rate or 802.11 HR) for transmissions of up to 11

Mbps. Global regulatory bodies and vendor alliances have endorsed this new high-rate standard, which promises to open new markets for WLANs in environments ranging from large enterprises to small offices and homes. With 802.11b, WLANs will be able to achieve wireless performance and throughput comparable to wired Ethernet.

Outside of the standards bodies, wireless industry leaders have united to form the Wireless Ethernet Compatibility Alliance. WECA's mission is to certify cross-vendor interoperability and compatibility of IEEE 802.11b wireless networking products and to promote that standard for the enterprise, the small business, and the home. Members include WLAN semiconductor manufacturers, WLAN providers, computer system vendors, and software makers such as 3Com, Aironet, Apple, Breezecom, Cabletron, Compaq, Dell, Fujitsu, IBM, Intersil, Lucent Technologies, No Wires Needed, Nokia, Samsung, Symbol Technologies, Wayport, and Zoom.

CONSIDERATIONS FOR CHOOSING A WLAN

While the bulk of these last sections has described how 802.11b WLANs are alike, there are still many ways for WLAN vendors to differentiate themselves in the marketplace to affect a customer's purchasing decision. We cover some of these areas in this section.

EASE OF SETUP

To install a WLAN, one must install and configure access points and PC cards. The most important piece of this effort is proper placement of the access points. Access point placement is what ensures the coverage and performance required by the network design. Several features provide assistance in the installation process:

- **Site survey.** For complete WLANs employing a cellular architecture, proper placement of access points is best determined by performing a site survey, in which the person installing the WLAN can place access points and record signal strength and quality information while moving about the intended coverage area. While most vendors provide a site survey tool, these utilities vary in the amount and quality of information they provide, as well as in their logging and reporting capabilities.

- **Power over Ethernet.** Some vendors ship access points that can be powered over the Ethernet cable that connects the access point to the wired network. This is usually implemented by a piece of equipment in the wiring closet. This equipment takes in AC power and the data connection from the wired switch and then outputs DC power over unused wire pairs in the networking cable that runs between the module and the access point. This feature eliminates the need to run an AC power cable out to the access point (usually located on the wall or ceiling), making installation quicker and more affordable.

Easy-to-use NIC and access point configuration tools. Once the access points are installed, both access points and NICs must be configured for use. As with any technical product, the quality of the user interface determines the amount of time required to configure the network for operation. In addition, some vendors supply tools for bulk configuration of access points on the same network, greatly easing network setup. Finally, having a variety of methods to access the access point is helpful to ensure simple setup. Configuration options include telnet, Web-based configurators, or SNMP-based c configuration utilities over the Ethernet cable either from a wireless station or via a serial port built into the access point.

EASE OF MANAGEMENT

Because an 802.11 WLAN differs from standard 802.3 and 802.5 wired LANs only at OSI layers 1 and 2 (physical and data link), we should expect at least the same level of manageability from these products as we find for wired networking products. At a minimum, the products should come with SNMP 2 support so that they can be automatically discovered and managed using the same tools employed for wired LAN equipment. And we should assess carefully what can be controlled via the SNMP MIB. Some products measure and control a number of Ethernet and radio variables in the access point, while others provide only a basic Ethernet MIB.

Beyond SNMP, it is useful to be able to configure and probe access points via an easy-to-use interface like a Web browser. Some vendors have built Web servers into their access points for this reason. Finally, the ability to manage, configure, and upgrade access points in groups simplifies WLAN administration.

RANGE AND THROUGHPUT

802.11b WLANs communicate using radio waves because these waves can penetrate many indoor structures or can reflect around obstacles. WLAN throughput depends on several factors, including the number of users, microcell range, interference, multipath propagation, standards support, and hardware type. Of course, anything that affects data traffic on the wired portions of the LAN, such as latency and bottlenecks, will also affect the wireless portion.

When it comes to range, more is not always better. For example, if the network requirement is for high performance (5.5 Mbps or 11 Mbps) and complete coverage, long range at lower network speeds (1 Mbps and 2 Mbps) may make it difficult to employ a channel reuse pattern while maintaining high performance.

MOBILITY

While 802.11b defines how a station associates with access points, it does not define how access points track users as they roam about, either at layer 2 between two access points on the same subnet, or at layer 3 when the user crosses a router boundary between subnets.

The first issue is handled by vendor-specific, inter-access-point protocols, which vary in performance. If the protocol is not efficient, there is a chance of packets being lost as the user roams from access point to access point. Eventually WECA and the IEEE are likely to create standards in this area.

The second issue is handled by layer-3 roaming mechanisms. The most popular of these is Mobile IP (see Figure 12.6), which is currently known as RFC 2002 in the IETF Mobile IP has an access point assigned as the *home agent* for each user. Once a wireless station leaves the home area and enters

FIGURE 12.6 Mobile IP is supported with these PCMCIA cards

a new area, the new access point queries the station for its home agent. Once the home agent has been located, a packet forwarding is established automatically between the two access points to ensure that the user's IP address is preserved and that the user can transparently receive his or her data. Since this protocol is not finalized, vendors may provide their own protocols using similar techniques to ensure that IP traffic follows a user across networks separated by a router (e.g., across multiple buildings).

An incomplete but useful alternative solution to the layer-3 roaming problem is to implement the DHCP across the network. DHCP enables any user who shuts down or suspends his portable computer before crossing to a new network to automatically obtain a new IP address upon resuming or turning on his notebook back on.

POWER MANAGEMENT

End-user wireless products are typically designed to work completely untethered, via battery power. The 802.11b standard incorporates Power Saving Protocol to maximize the battery life of products using wireless devices.

SAFETY

As with other wireless technologies, WLANs must meet stringent government and industry standards for safety. There have been concerns raised regarding the health risks to people who use wireless devises. To date, scientific studies have been unable to attribute adverse health effects to WLAN transmissions. In addition, the output power of WLAN systems is limited by FCC regulations to under 100 mW, much less than that of a mobile phone, and any health effects related to radio transmissions most likely would be correlated to power and physical proximity to the transmitter.

SECURITY

The WEP 40-bit encryption built into 802.11b WLANs should be sufficient for most applications. However, WLAN security must be integrated into an overall network security strategy. In particular, a user may implement network-layer encryption such as IPSec across both wired and wireless portions of the network, eliminating the need to have 802.11 security in place. Or customers may choose to have critical applications encrypt their own data, thereby ensuring that all network data such as IP and MAC addresses are encrypted along with the data payload.

Other access control techniques are available in addition to the 802.11 WEP authentication technique. For one, there is an identification value called an ESSID programmed into each access point to identify which subnet it is on. This can be used as an authentication check—if a station does not know this value, it is not allowed to associate with the access point. In addition, some vendors provide for a table of MAC addresses in an Access Control List to be included in the access point, restricting access to clients whose MAC addresses are on the list. Clients can thus be explicitly included (or excluded) at will.

COST

Hardware costs include adding access points to the network infrastructure and WLAN adapter cards to all wireless devices and computers. The number of access points depends on the coverage area, number of users, and types of services needed. The coverage area of each access point extends outward in a radius. Access point *zones* often overlap to ensure seamless coverage. Clearly, hardware costs will depend on such factors as performance requirements, coverage requirements, and vendor product range at different data rates.

Beyond equipment costs, a customer must take into account installation and maintenance expense, including the costs of poor product quality (help-desk support costs, end-user productivity). These costs can dwarf the initial equipment costs of a WLAN. Products that are simple to install, use, and manage, and that perform up to their specifications, may be worth significantly higher initial equipment investment. Features mentioned earlier, such as power over Ethernet, bulk configuration of access points, and a rich set of management tools, will lower the overall cost of a WLAN.

SUMMARY

802.11 WLANs already are commonly used in several large vertical markets. The 802.11b standard is the first standard to make WLANs usable in the general workplace by providing robust and reliable 11-Mbps performance, five times faster than the original standard. The new standard will also give WLAN customers the freedom to choose flexible, interoperable solutions from multiple vendors, since it has been endorsed by most major networking and PC vendors. Broad manufacturer acceptance and certifiable interoperability means users can expect to see affordable high-speed wireless solutions proliferate throughout the large enterprise, small business, and home markets. This global WLAN standard opens exciting new opportunities to expand the potential of network computing.

ENDNOTE

Report on Wireless Technologies, Frost & Sullivan, 2000.

APPENDIX A

Sun Microsystem's View of the Connected Home

(Powered by Java Embedded Server)

EXECUTIVE SUMMARY

The pervasiveness of the Internet is enabling a connected lifestyle. The phone dial tone is giving way to the Internet Webtone, which will bring exciting new applications to consumers wherever they may be: at home, in cars, on cell phones, or in hotel rooms. In the home, the key to this connected lifestyle is the home gateway, a black box that connects devices and appliances in the home to each other and to the Internet.

Broad opportunities exist for businesses—service developers, service providers, and device manufacturers—to deliver exciting new services to the networked home through the home gateway. Open standards, such as the Open Services Gateway initiative (OSGi) for the network delivery of managed services, and industry

alliances such as the Internet Home Alliance, are critical for interoperability and consumer adoption of these technologies. Sun is building end-to-end solutions for this market, including Java Embedded Server software—an ideal solution for powering residential gateway devices to enable dynamic delivery of network services on demand.

THE DOT COM HOME

The Internet is enabling the new connected lifestyle. Already, many of us communicate, shop, perform research, and do other business on the Internet, helping to create the new Internet economy. As the Internet extends its reach down to more and more devices in our homes and cars as well as our businesses, we can embrace this technology in a whole range of ways to make our lives easier and more productive. Powerful new services for communication, entertainment, home control, and information will enable us to stay connected no matter where we are.

Imagine a home in which all the different devices are connected. Consumers find out about an early meeting for the following day. When scheduling the meeting into a calendar program at the office or a handheld PDA, they also reset their home alarm clocks to wake them an hour early. Automatically, the alarm clock resets furnaces and coffeemakers to turn on 15 minutes before they need to get up. It sets the lights to come on as soon as the alarm goes off. It even resets the landscaping sprinklers so they don't come on while the person is in the shower, reducing the water pressure.

Expand the picture to include audiovisual equipment, security and health monitoring systems, kitchen appliances, pools, and spas. Telephone services, entertainment and information access, and e-commerce services all come into play, and an optional wireless Web pad to access provides the services from anywhere in the house. The consumer can take advantage of these services at low incremental fees, in the same way that phone companies offer call waiting and conferencing services to enhance basic phone service.

The service-driven network is enabling the Internet lifestyle, giving consumers access to a host of exciting new services and applications on any device, anywhere, anytime. The service-driven network gives consumers the same degree of dial tone reliability that traditional phone networks provide, while delivering new services

that reside on the network. This new computing architecture has three primary components:

- Access—various network-enabled devices

- Delivery—personalized value-added delivery of Java

- Resources—the back-end data center servers, storage, and software that act as the computing engine and warehouse of information

- Network Services Clients

- Network Infrastructure Service

The connected home is a key market segment of the service-driven network. Electricity, gas, phone service, cable TV—these utilities have been around for a while. The next utility is the Internet. Soon the Internet will be everywhere: in the car, on the cell phone, and in the home. In tomorrow's home—Sun's connected home—networked home appliances will be as commonplace as the telephone is today. Microprocessor-based consumer products are already common in our daily lives: alarm clocks, coffee machines, televisions, cars, air conditioners, and phones. Stereo systems, kitchen appliances, climate-control systems, utility meters, and security systems are just a few of the many devices that are now being connected to each other and the Internet, offering consumers powerful new ways to use technology to enhance their daily lives. Coupled with the broadband pipe to the home, the connected home enables service providers to deliver exciting new services such as audio/video on demand, dial tone on demand, unified mailbox for e-mail/voicemail/fax, home security, and energy management to the various networked devices in the home.

Sun offers end-to-end solutions to support the opportunities for an enhanced lifestyle that are being created through the pervasiveness of the Internet. Put another way, the Internet is a key driver to the connected home. This is a key market to drive these services, and Sun can bring reliable Webtone for these next-generation services to the home.

MARKET TRENDS

Several important trends are converging today to create an environment in which the connected home is inevitable:

- Rapid growth of broadband

- More than 12 million homes in the U.S. alone will have broadband access by 2002 (IDC/Dataquest).

- 10 million homes will have broadband access by 2003 (Yankee Group).

- U.S. broadband revenue will increase from $2.4 to $14.8 billion by 2005 (Forrester Research).

- Emergence of network-enabled devices

- Worldwide information appliance shipments will grow from $11 million to $89 million annually by 2004 (IDC).

- The home networking equipment and residential gateway market will grow from $600 million to $5.7 billion by 2004 (Cahner's In-Stat).

- Emergence of local networks: more than 20 million homes will have an in-home network by 2003 (Parks Associates).

- A move to service-driven networks

- Broadcast television is moving toward pay-per-view, as well as dial tone on demand.

- Activities such as using a mobile phone to remotely access home security or audiovisual systems, and using the smart card in the phone for storing personal information or to secure transactions, will soon be everyday occurrences.

CONSUMER ACCEPTANCE

- 21 million U.S. households have an interest in the concept of the networked home.

- 42 percent of households with PCs would consider using a network to enhance communication across their families.

- About 12 million are likely to install such services over the next 12 months.

- 39 percent would use such services to enable the viewing of downloaded video anywhere in the house.

- 37 percent would use them to monitor and control the heating systems.

- 33 percent would use them to listen to music anywhere.

- 36 percent would use them to run household appliances.

NEXT-GENERATION INTERNET SERVICES

Deregulation of the telephone, cable, and utility industries has blurred the roles of these industries. Cable operators are moving to offer Internet access and phone services in addition to basic cable services. Phone companies are offering high-speed Internet access and eventually will offer video-on-demand services. Utilities are exploring energy management as well as the potential of becoming Internet service providers.

In this competitive environment, these service providers must not only protect existing revenue streams, but also generate new revenue. They are looking increasingly at the Internet to offer new value-added services to the networked home customers. Some of the new services in development for the networked home include communication, entertainment, home control, and information services.

COMMUNICATION SERVICES

Voiceover IP (VoIP) and unified messaging services are just a few of the myriad ways to bring telephony services to the home. VoIP supports directory service standards to help users locate other users, the use of touchtone signals for automatic call distribution and multiple phone lines on demand, individual mailboxes, conference calling, call transfer—all the services users already expect at the office are now available at home. Another major advantage of VoIP and Internet telephony is the avoidance of tolls charged by ordinary telephone service. Unified messaging services change the way users think about communicating, by letting the user receive any form of message in an in-box and collect it via any method: phone, voicemail, fax, e-mail, Web page, or answering service. Cell phones can act as transponders, allowing busy parents to track their children's whereabouts. The successors to today's popular chat rooms and instant messaging may be accessed.

Audio- and video-on-demand services and digital video recorders give the consumer access to a tremendous variety of specialized entertainment programming that can be started, stopped, and paused at will—all from the Web pad, without phoning or going anywhere. Consumers will have the ability to play movies from the DVD player in the den or on demand from a service provider, to any television in the home.

HOME CONTROL SERVICES

Home security, energy management, and home automation free consumers from worry and tedious routine tasks. Device monitoring services can run remote diagnostic tests to ensure that appliances and systems are operating properly. They can also receive alerts from devices indicating a functional problem. Many problems can be resolved remotely, or if needed, service agents can be automatically dispatched to the home to troubleshoot in person. Device integration services provide the means for appliances and other devices to communicate with each other to synchronize operations, such as the alarm clock telling the coffeemaker to turn on. Home network management supports multiple PCs and many other devices that share a broadband connection, and peripherals like printers. The consumer need not know anything about proxies, firewalls, and DHCP servers—the home networking service provider takes care of everything.

INFORMATION SERVICES

Integrated and personalized e-commerce services will greatly enhance the consumer shopping experience. The consumer will scan product bar codes and add items to an online shopping list as stocks run low. The ingredients for a recipe shown on a TV cooking program can be added by simply clicking a button on the remote. The shopping list is transparently converted to an order with an online grocer and delivered to the doorstep. Manufacturers will be able to provide targeted promotions based on a specific consumer's consumption profile (if the consumer wishes to receive them), as opposed to barraging customers with spam and junk mail. In addition, these personal profiles will customize homes, cars, and even hotel rooms to specific consumer preferences when an individual enters them using a smart card, or automatically make airline, hotel, and car reservations (based on the defined preference) as soon as a new trip is entered on an on-line family calendar.

Several studies have shown that consumers are willing to accept more technology if it adds real value, is similar to existing approaches, and is relatively transparent to them (cell phones are examples), but they are not interested in having to learn and implement these new technologies themselves. Consumers won't want to buy a device integration system, but are likely to be attracted by a home entertainment or home security service that works via the Internet to seamlessly provide services they can access from anywhere—as long as the technology to make this happen comes along with the service.

THE HOME GATEWAY IS KEY

So what will make all this possible? Obviously, just putting a wired PC in the home is not the answer—the difficulties of maintaining it and the technical knowledge required to do so could outweigh the value of the services obtained. What will make these services attractive to the consumer is a dedicated, small device, connected to the rest of the house, that requires minimal or no upkeep on the part of the homeowner.

Thus, the real key to the connected home is the home gateway, the nerve center of the networked home.

The connected home of tomorrow will connect all of the networks that already exist in the home—electrical, telephone, wireless—and then connect each one with any number of external networks via the Internet. The box that will network appliances within the home and connect them to the Internet is the home gateway. This type of network enables washing machines to download new washing programs dynamically, electronic toys to download updated game programs, and the consumer to turn off the oven, iron, or lights remotely after leaving for a trip. Most importantly to the consumer, all the services of the home gateway can be managed by external service providers. Just like the cable or phone service, home gateway services are just there for the consumer when they are needed—the consumer does not need to understand anything about how or why they work, just that they do. While home gateways are just one application of open services gateways (the software is expected to run in everything from gas pumps to soda machines to power substations), the home is an important and

WHAT A HOME GATEWAY IS

What is a home gateway? It could be a:

- Cable modem
- Set-top box
- DSL modem
- Web phone
- Dedicated residential gateway device

The specialized hardware and software required for a gateway can be built into a new, specialized device or embedded into an existing device. In effect, adding an embedded server—a special-purpose, low-memory, software server (not a Web server)—to any broadband termination device, transforms it into a home gateway.

The look and feel of the home gateway device will be influenced by the business model of the service provider who delivers it. A cable operator would embed this functionality in a cable modem or set-top box device. A telephone company might choose to include this functionality in a DSL modem or router, and a utility company could target the residential utility meter. To the consumer, the residential gateway will be a black box that is not even visible in the home and that is managed by the service provider.

Why it is Necessary

A home gateway links devices in local networks in the home to the Internet and external service providers, creating a focal point for enterprises and service providers to deliver services to client devices. The home gateway serves two primary roles:

- As a hub to connect and manage the intelligent appliances in the home

- As a communications gateway between the home and the outside world

Many types of devices can already communicate with each other. Why do we need another device to connect them? Even though a lot of important technology has already been developed to address many of these issues, all the infrastructure is not yet in place to make these connections seamless. Among the most important elements of this infrastructure is home wiring. New and existing houses need to be wired with copper, coaxial cable, fiber, or some other alternative to receive external network services. Once the homes are connected to the outside world, the existing telephone or power lines, radio, wireless, infrared, or some other technology can be used to create secure internal networks. Current service providers must learn about this new technology and develop their own infrastructures to serve customers, and new providers must come into being to provide services that today may not even be imagined.

How It Works

The home gateway is an embedded server that is inserted into the network to connect the external Internet to internal clients. The gateway is inserted between the service provider's network and the home (or the remote branch office), LAN, and client devices. The gateway separates the topology into the external and internal networks. Services are delivered from trusted service providers on the external network to the gateway or internal clients. The gateway is typically a zero-administration device that is secure. It functions as a bridge between the internal and external components.

Business Models

Now that we know what can happen, the question is how to make it happen. The current industry landscape is very fragmented, with many products, technologies, and standards. Sun is one of an increasing number of companies developing products and services for the connected home, and it is clear that no one company can create everything that is needed to enable the Internet lifestyle. Certainly one important driver for this development is the enhanced lifestyle it can bring to the consumers who will buy these products and services, and, in turn, the revenues it will mean for the creators and builders of these new technologies. But, another important reason is that both large and small companies are testing the boundaries of the corporate market that has been their lifeblood and are finding that they need to expand into the consumer market.

"It's a push market trying to create a pull market," said Michael Wolf, director of residential technologies at Cahners In-Stat Group, as quoted in *The Industry Standard* in October 2000. "You need infrastructure for these services, but there won't be demand for infrastructure until there are compelling services."

What are some of the roles that service and infrastructure providers can play in creating the Internet lifestyle?

Roles That Enable the Connected Home

Because of the vast new opportunities it offers, the potential of the home gateway is exciting not only to consumers but also for service providers, gateway operators, and vendors of hardware, software, and devices. Several roles are involved in the end-to-end deployment of a home gateway, and any individual company may perform one or many roles.

Service Provider

The service provider delivers a service or content that is of benefit to the consumer. The service is an application that is downloaded by the managed services provider into the gateway device. In the gateway model, service providers can focus on developing content, and outsource delivery and management. They rapidly deploy new services and generate service revenue streams. Providers may lower their costs by sharing a gateway with multiple providers and administering systems remotely. Service providers may choose to emulate the cellular phone model, in which the consumer gets the gateway device free—and revenue is obtained through monthly service payments.

Managed Services Provider

The managed service provider operates, manages, and maintains the gateway devices and their applications. The gateway offers new business and revenue opportunities for the provider, attracting new customers as well as increasing customer retention by adding services for existing customers. Gateway services can also increase network traffic and revenue. Costs can be lowered with gateways that are vendor-neutral, and the Write Once, Run Anywhere concept lets the provider choose from more platform options. The managed services provider could be the telephone company, cable operator, ISP, or CLEC (competitive local exchange carrier). The role of the managed services provider is to:

- Download, remove, start, and stop an application
- Control the resources of the gateway, check the operating cycle of the gateway, and manage application versions
- Define and control access rights between the gateway and service providers
- Secure the communications between the gateway and service providers
- Check the validity and the rights of any services that are dependent on other services

Network Access Provider or Service Aggregator

The network access provider provides and manages the network used to access the gateway. This provider could potentially be a telephone company, cable operator, or ISP.

Hardware, Software, and Device Vendors

For hardware, software, and device vendors, the home gateway offers access to a much larger market, increased revenues due to standardization, and the ability to create solutions on a vendor-neutral platform, which also reduces the risks inherent in proprietary solutions. The Write Once, Run Anywhere concept helps to reduce software development costs.

Gateway Retailer

If the gateway is not provided free of charge by a service provider or managed services provider, the gateway retailer offers it for purchase by the consumer at a retail outlet, an additional new line of revenue.

BUILDING THE DOT COM HOME WITH JAVA EMBEDDED SERVER SOFTWARE

Ultimately the consumer gains tremendous benefits from the home gateway. The consumer can obtain better availability of more services, at a lower cost because just one gateway delivers multiple services. One-stop service shopping offers unparalleled convenience for today's busy consumer, who can deal with several

WANs or LANs at the same time, switch easily between service providers to get better services or a lower price, and securely access services through a service aggregator.

Technology Standards

A number of standards, alliances, and programming models have arisen to help bring some unity to the fragmented industry that is evolving the connected home.

BROADBAND STANDARDS

Digital Subscriber Line (DSL)—DSL is a technology for bringing high-bandwidth information to homes and small businesses over ordinary copper telephone lines.

Data Over Cable Systems Interface Specifications (DOCSIS)—Now known as CableLabs Certified Cable Modems, DOCSIS is an interface standard for cable modems, the devices that handle incoming and outgoing data signals between a cable TV operator and a personal or business computer or television set.

Wideband Code Division Multiple Access (WCDMA)—WCDMA is a third-generation mobile wireless technology offering very high data speeds to mobile and portable wireless devices. WCDMA is an ITU standard derived from code-division multiple access (CDMA) that is officially known as IMT-2000 direct spread.

HOME/LOCAL NETWORK STANDARDS

Phone Line—HomePNA is an industry standard for interconnecting computers within a home using existing telephone lines and registered jacks in a transmission technology similar to traditional Ethernet. Using HomePNA, multiple computer users in a home can share a single Internet connection, open or copy files from different computers, share printers, and play multiuser computer games.

Wireless—Bluetooth is a computing and telecommunications industry specification that describes how mobile phones, computers, and personal digital assistants (PDAs) can easily interconnect with each other and with home and business phones and computers using a short-range, wireless connection. The HomeRF standards body has developed the Shared Wireless Access Protocol (SWAP) specification as a wireless LAN solution that operates in the 2.4 GHz band at a raw data rate of 1.6 Mbps. IEEE 802.11b, also known as 802.11 High Rate (HR) and wireless Ethernet, is a high-speed, wireless Ethernet connectivity standard designed for both enterprise LAN environments and the home network. Operating in the 2.4 GHz band, 802.11b is fast, scalable, and has an impressive range: it operates at a raw data rate of 11 Mbps, can accommodate up to 128 nodes on a network, and has a range of up to 150 meters.

Power Line—Echelon's LonWorks system is an open, networked automation and control solution for the building, industrial equipment, transportation, and home markets. Based on physical transceivers and application layer software, LonWorks nodes can be connected on multiple types of media; twisted pair and power lines are the most common.

CEBus (Consumer Electronic Bus) also known as EIA-600, is a standard for powerline networking using spread-spectrum technology. The CEBus Industry Council is merging the CEBus protocol and HomePnP into the SCP standard. Simple Control Protocol (SCP) is a networking technology for devices with limited memory and processing power and networks with low bandwidth. Devices that would benefit from SCP include lights, home security devices, home automation devices, and other small appliances that are not able to support TCP/IP networking, or that connect to a home network through a low-speed powerline medium.

The HomePlug Powerline Alliance, formed in April 2000, has announced a powerline technology for home networking that will support a raw data rate of 14 Mbps. The first version of the HomePlug specification is expected early next year.

OPEN SERVICES GATEWAY INITIATIVE (OSGI)

To fully develop the market for residential gateways of a broadly embraced open standard is critical. The Open Services Gateway initiative (OSGi), is an open industry effort founded in March 1999 by 15 technology companies, with the objective of providing a forum for developing open specifications to deliver multiple services over wide area networks to local networks and devices. The Initiative is also focusing on accelerating the demand for products and services based on those specifications worldwide through the sponsorship of market and user education programs. Today, more than 80 companies have committed to support the full incorporation and charter of the organization. Sun is a founding member of OSGi and a major contributor to its technical foundation.

The four major OSGI premises are:

- The networked home is the next frontier.

- The Internet and new technologies enable new services, value chains, and business models.

- Standards are required for the market to take off.

- Consumers need an end-to-end solution that requires Internet connectivity for home equipment. This connectivity is provided by a gateway.

The OSG specification is the missing link in the networked delivery of managed services from broadband network to local networks in the home. Java technology provides the flexibility to support the wide range of phone line, powerline, and wireless network standards.

The specification is a layer framework application based on Java technology. It gives service providers, network operators, device makers, and appliance manufacturers the vendor-neutral application and device layer APIs and functions they need. These APIs define a set of core and optional APIs that together define an OSG compliant gateway. The OSGI specification includes APIs for service cradle-to-grave lifecycle management, interservice dependencies, data management, device management, client access, resource management, and security. Using these APIs, clients load network-based services on demand from the service provider, and the gateway manages the installation, versioning, and configuration of these services. In addition to the APIs, the OSG specification defines a number of required or optional services, such as Web server, alarms, logging, data management, and more. For additional information, see www.osgi.org.

INDUSTRY ALLIANCES

INTERNET HOME ALLIANCE

The Internet Home Alliance was formed in 2000 by twelve founding member companies, with a vision of enabling and accelerating the development of the Internet lifestyle. Sun is a founding member of the Internet Home Alliance, along with 3COM, Best Buy, Cisco, General Motors, Honeywell, Invensys, Motorola, New Power Company, Panasonic, and Sears. The Internet Home Alliance seeks to enable a world of ubiquitous products and services that are as commonly accepted as the telephone and television are today.

The Internet Home Alliance is a facilitator for development of the Internet home market. The organization acts as an influencer to manufacturers and service providers, channel partners, and customers. The organization should be a source of stability and a builder of primary demand within the consumer community.

The Internet Home Alliance was conceived to accomplish two key goals:

- Reduce customer confusion through simple and affordable solutions

- Educate customers on the value and availability of Internet Lifestyle solutions

No single company can accomplish these goals, so the Internet Home Alliance was formed to combine critical skills, resources, and relationships from many companies that will enable the Internet home market to flourish. Additional members will be recruited to provide the right talent mix, including solutions partners, home integrators, service providers, and support organizations.

SUN PRODUCTS AND TECHNOLOGIES FOR THE CONNECTED HOME

GATEWAY

Sun's Java Embedded Server software is an ideal solution for powering residential gateway devices to enable dynamic delivery of network services on demand. The Java Embedded Server product is a small-footprint application server that can be embedded in a networked device—like a home gateway—and provides lifecycle management of services deployed to the device over the network. It allows development, deployment, and installation of applications and services to embedded devices on a just-in-time basis. Using Java Embedded Server software, companies can quickly respond to changing market requirements by installing and managing new software and services on the devices over the network—dynamically and securely. Java Embedded Server software is written to the Java platform, so it runs on a wide variety of devices. The software has two primary components:

- **Framework.** The Java Embedded Server framework is small enough to fit within the footprint of nearly any target device, and is composed of APIs for lifecycle management of plug-n-play services and applications. This modular model provides installation, versioning, content management, monitoring, service discovery, and more.

- **Services.** Plugging into the framework are servlet-like objects called services, which represent the modular services that can be deployed over the Internet, including HTTP, logging, thread management, remote administration, and servlet support. To provide and manage services for consumer devices around the home, optional Services can be loaded into the framework on demand. These optional bundles could handle everything from phone dial tone on demand to home security services to live pay-per-day sports scores or other services.

SUN PRODUCTS AND TECHNOLOGIES FOR THE CONNECTED HOME

The product's small size, architecture, use of Java technology, and extensibility are targeted primarily at developers and manufacturers of residential gateway products.

The features of Java Embedded Server 2.0 include:

- Thin-server framework for dynamic delivery of network services

- Designed to be fully compliant with the OSGi 1.0 specification (HTTP, log, and device access services)

- Comprehensive Developer Tool kit, including centralized tools portal with plug-in for Forte for Java

- Community Edition software to expedite development of OSGi-compliant applications and services

- User Portal service to automatically provide a user interface on a graphical device (like a Web pad) for interaction with services running on the gateway device HTML-based administration console for gateway management

- Support for PersonalJava 3.0.2 software on VxWorks

For more information on Java Embedded Server software, visit www.sun.com/

OPENING THE DOOR TO THE CONNECTED

Sun's home gateway software bridges all of the devices on the home's network with external networks and their services. It provides a focal point for service providers to deliver value-added services to the networked devices in the home. Scott McNealy, CEO of Sun Microsystems, recently remarked:

"Sun won't get into the business of creating the connected home, but we'll certainly provide much of the technology, the same technology that also powers the Net. After all, if you look at the connected home, you'll realize that it uses the same backbone—open Internet standards and cross-platform technologies—used to connect the Net to cars, offices, governments, and just about anything else. This is an important point, because the connected home must speak the same language as the rest of the Net."

Sun's consumer and embedded technologies are being adopted in a variety of exciting markets. In addition to the home gateway market, these markets include wireless communications, digital interactive TV, and automotive. Each of these markets is characterized by a movement towards providing value-added services over a network—and Java Embedded Server technology, as well as Sun's other consumer and embedded technologies, play a key role in making this happen. Opportunities also abound for manufacturers of devices needing lifecycle management that are capable of hosting Java technology; examples include small-business servers, printers, copiers, and point-of-sale (POS) devices such as vending machines, gas pumps, kiosks, and cash registers. Through partnerships with service providers, device manufacturers, content providers, and system integrators, Sun is committed to providing end-to-end solutions based on open standards that will make consumers' lives more productive and enjoyable.

APPENDIX B

USING THE INTERNET CONNECTION WIZARD IN MICROSOFT SMALL BUSINESS SERVER

INTRODUCTION

This document describes Internet connectivity with the Microsoft BackOffice Small Business Server family version 4.5 and introduces the Small Business Server Internet Connection Wizard (ICW). This document includes a detailed walkthrough that explains the configurations made by the wizard to help you determine whether the settings are configured appropriately for your network environment.

Before delving into the Small Business Server Internet Connection Wizard, you will want to understand how Small Business Server is connected to the Internet, because the type of connection affects the choices you'll make.

Small Business Server supports many types of connectivity hardware, which can be grouped into three categories: dial-up connections, routers, and broadband devices.

Dial-up Configuration

A dial-up connection is established using Dial-up Networking or Remote Access Services (RAS). The hardware used for a dial-up connection is typically an analog modem or an ISDN terminal adapter. These devices must be configured through the Modems page of Control Panel. They may be connected to Small Business Server through a serial port or a multiport serial board, or they may be internal.

Router Configuration

Two configurations, or network topologies, are commonly used to connect to the Internet through a router. The Small Business Server ICW supports both configurations.

In the first topology the router is connected directly to the local area network (LAN). The router's internal Internet protocol (IP) address is set as the default gateway for each computer on the LAN, so the router handles clients' requests for resources outside the available local addresses (addresses beyond the LAN, on the Internet). This topology requires every client on the LAN to have a valid IP address unless the router is doing network address translation (NAT).

In the second topology the router connects to the LAN through a second network adapter installed in the Small Business Server-based computer. The server's first network adapter connects the server to the LAN, and the second adapter connects the server to the router, which in turn connects to the Internet. The server's TCP/IP configuration has no default gateway on the internal network adapter, but the default gateway is configured on the external network adapter. No default gateway is configured on the clients because they will connect to the Internet through the Microsoft Proxy Server firewall.

The second topology is more secure because it takes advantage of the Proxy Server firewall features in Small Business Server. All incoming and outgoing packets must go through Small Business Server before being routed to their intended destination, so Proxy Server can monitor the packets and prevent outsiders from reaching your LAN.

Note: Some routers can act as Dynamic Host Configuration Protocol (DHCP) servers. If your router has this feature, it must be disabled to prevent conflicts with the DHCP service running on Small Business Server. The DHCP service included with Small Business Server has a rogue detection feature that will shut down the service if another DHCP server is detected. If the router is also a DHCP server, clients will not be able to receive DHCP addresses from Small Business Server. For more information about this DHCP feature, see http://support.microsoft.com/support/kb/articles/q177/6/10.asp.

FULL-TIME/BROADBAND CONNECTIONS: ADSL, CABLE MODEMS, AND MMDS

Broadband connections are becoming more feasible for small businesses because costs associated with them are becoming more affordable. A second network adapter is required for a full-time/broadband connection, which makes it similar to the second topology. This topology is secure because all inbound and outbound Internet requests are first filtered by the Proxy Server, then delivered accordingly. When using a full-time/broadband connection to the Internet, you must have a static IP address, supplied by your ISP, for your broadband connection on the external network adapter.

Broadband Configuration. The LAN is connected to one network adapter, and the other network adapter is connected to the broadband device.

SMALL BUSINESS SERVER INTERNET CONNECTION WIZARD WALKTHROUGH

The following walkthrough introduces the Small Business Server Internet Connection Wizard (ICW), explains the questions that the wizard asks, and helps you understand what answers to give in order to correctly configure the Internet connectivity settings in BackOffice Small Business Server.

Information Needed to Run the Wizard

The information you need to run the ICW depends on the Small Business Server–based computer's connection to the Internet.

Note: A printable form that includes all the information for each type of connection is available on the Configure Hardware page of the Internet Connection Wizard and in the "References" section of this document.

Dial-up Connection

For a modem connection, or any dial-up connection that uses a phonebook entry to dial, you will need the following information from your Internet service provider (ISP):

- The account name (or UserID required to dial into your ISP)
- The password for the account
- The type of e-mail used: Microsoft Exchange Server (which uses SMTP, or Simple Mail Transfer Protocol) or POP3
    - For Exchange Server (SMTP) e-mail, you will also need the host name or IP address of the ISP's SMTP server. If your ISP queues mail for Small Business Server to pick up when it dials in, you will need to know the appropriate signaling, or "de-queuing" command.
    - For POP3 mail, you will need the host name or IP address of the POP3 and SMTP server at your ISP.
- Your Internet domain name

Router Connection

To correctly configure Small Business Server for Internet connectivity through a router, you will need the following information:

- The type of e-mail used: Exchange Server (SMTP) or POP3

    - For Exchange Server (SMTP) e-mail, you will also need the host name or IP address of the ISP's SMTP server. If your ISP queues mail for Small Business Server to pick up when it dials in, you will need to know the appropriate signaling, or de-queuing command.

    - For POP3 mail, you will need the host name or IP address of the POP3 and SMTP server at your ISP.

- Your Internet domain name

- The IP address of your router

- The IP address of the Domain Name Servers (DNS) for your ISP

- Whether the router dials on demand or not

Broadband/Full-time Connection

A broadband connection requires a second network adapter configured with a static IP address, which must be configured before using the wizard, and:

- The type of e-mail used: Exchange Server (SMTP) or POP3

    - For Exchange Server (SMTP) e-mail, you will also need the host name or IP address of the ISP's SMTP server. If your ISP queues mail for Small Business Server to pick up when it dials in, you will need to know the appropriate signaling, or de-queuing command.

    - For POP3 mail, you will need the host name or IP address of the POP3 and SMTP server at your ISP.

- Your Internet domain name

- The IP address of your router

- The IP address of the DNS servers for your ISP

- Whether the router dials on demand or not

Optional Information (Web Publishing Wizard)

The following information is required if you choose to edit your Web site through the Web Publishing Wizard included with Small Business Server instead of through the Microsoft FrontPage Web site creation and management tool or another Web site editing program:

- Your Web page URL. This is the Web page URL that is publicly available on the Internet.

- Your Web posting URL. This is the URL on the Internet through which you make changes to the files of your Web site and where you update the data files that make up your Web site.

- Your Web posting account name and password. This is the account name and password required to make changes to your Web site.

Once you have the information listed above, you are ready to run the wizard. The remainder of this section explains each of the pages the wizard displays. Note that the values you enter in the wizard determine which pages are presented. The "References" section includes flowcharts that explain how your choices affect the pages you see. Do not expect the wizard to show you all of the pages covered in this section.

Summary of Pages

The Small Business Server Internet Connection Wizard (ICW) can be started from the Small Business Server Console in a couple of ways. ICW can be called from the To Do List or from Manage Internet Access under More Tasks. If Internet connectivity changes need to be made after the wizard has been run, select Configure Internet Hardware from the Manage Internet Access page of the Small Business Server Console, and then make any changes necessary.

The Welcome to the Small Business Server Internet Connection Wizard Page

This first page is a simple introduction. Click Next.

APPENDIX B: USING THE INTERNET CONNECTION WIZARD

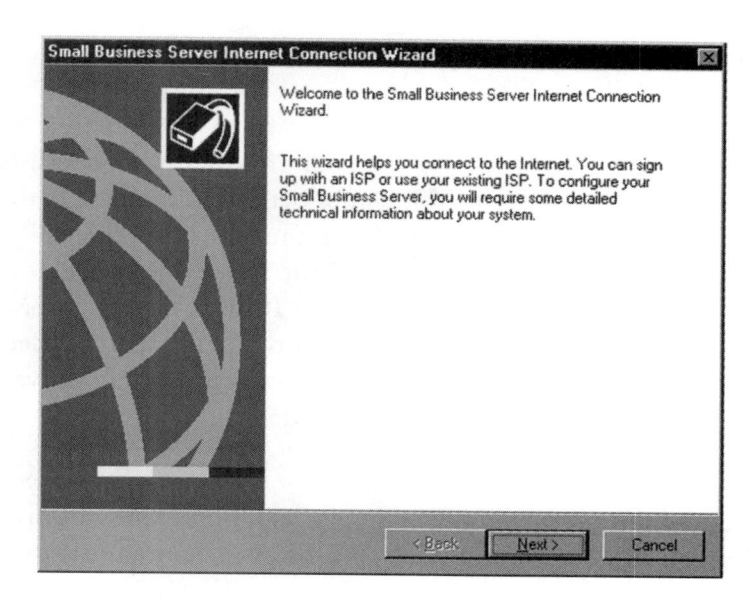

FIGURE 1 The Welcome to the Small Business Server Internet Connection Wizard page.

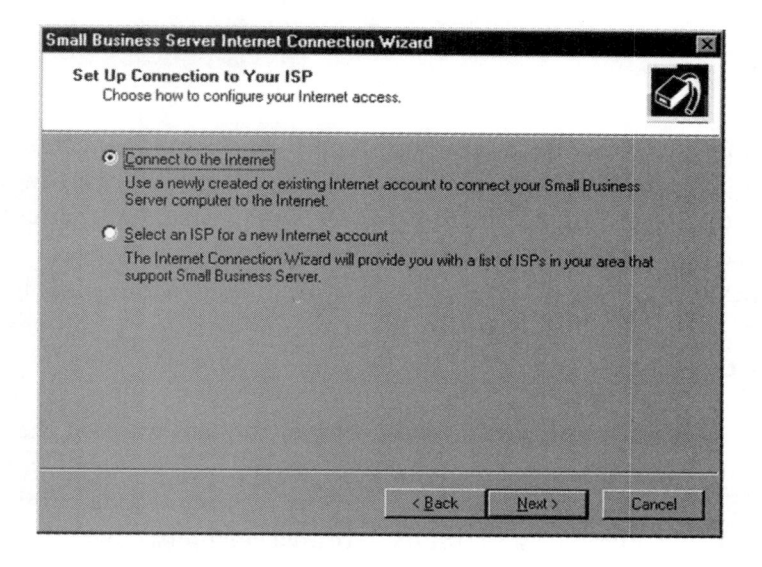

FIGURE 2 The Set Up Connection to Your ISP page.

The Set Up Connection to Your ISP Page

Select an ISP for a new Internet account is used to connect to a Microsoft Referral Server that helps locate ISPs in your area that have special offerings for customers of BackOffice Small Business Server. In some cases these ISPs provide an online sign-up that configures the appropriate Small Business Server services. By selecting Select an ISP for a new Internet account, you launch the Internet Connection Wizard. The ICW then dials the Microsoft Referral Server and downloads a list of ISPs that have special offerings for Small Business Server customers. Therefore, in order to take advantage of this feature, you must have a functional modem on your server.

If it supports online signup, once your new ISP has received your user, system, and billing information, the signup server generates an .ins file that is downloaded to your server. This .ins file will configure your dial-up networking connection, Proxy Server, Microsoft Exchange Server, and the information required by the Web Publishing Wizard. Some ISPs distribute sign-up floppy disks. If your ISP does this and is configured specifically for Small Business Server, you will not need to use the Small Business Server Internet Connection Wizard. Follow the instructions included with the sign-up disk to continue the installation.

The Configure Hardware Page

If you choose Connect to the Internet from the Set Up Connection to Your ISP page (Figure 2), you are presented with the Configure Hardware page. Through this page you tell Small Business Server what Internet communications hardware connects the server to your ISP. The type of hardware determines what other information you'll need later in the wizard; clicking the Form button for each option displays the information you'll need if you select that option.

Select Modem or Terminal Adapter whenever the hardware on the server is detected by the Windows NT operating system as a modem and can be accessed through dial-up networking and the Remote Access Service. (This includes analog modems and ISDN terminal adapters.) Before beginning with the Internet Connection Wizard, these devices should be installed

APPENDIX B: USING THE INTERNET CONNECTION WIZARD

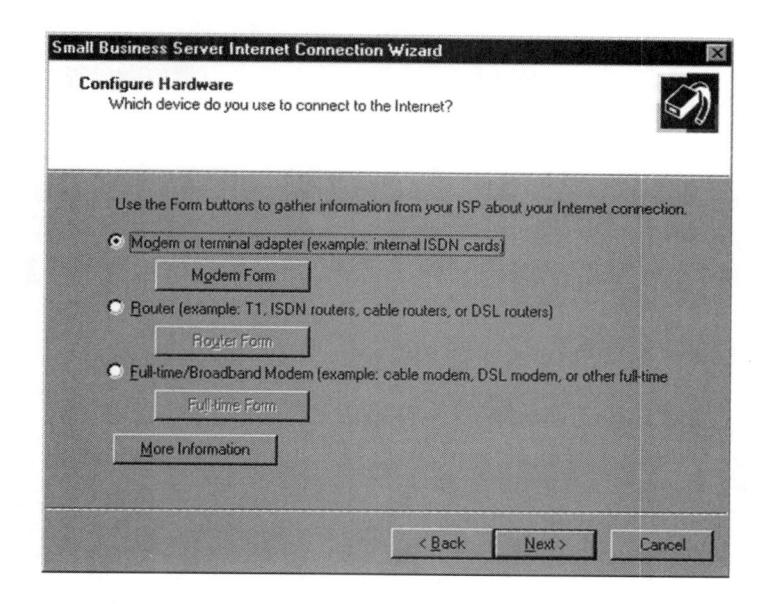

FIGURE 3 The Configure Hardware page.

through the Modems area of Control Panel and be configured as ports through the Remote Access Service with dial-out access.

Select Router if the connection to the ISP is through a separate router, such as Frame Relay, an ISDN router, a cable router, or a DSL router. These routers must be configured separately. Before proceeding, you will also need to gather the router information from the lists outlined in the "Information Needed to Run the Wizard" section. Selecting this option displays the Set Up Router Connection to ISP page.

The Full-time/Broadband Modem selection is designed for all connection methods that provide a full-time, high-speed connection to your ISP through a second network adapter. This connection is separate from the network adapter attached to the local network. These full-time connections include ADSL modems, cable modems, and the access to these devices through second network adapters. All of this hardware must be configured before proceeding with this wizard. Also, you will need to configure the second (external) network adapter with a static IP address before continuing with the wizard. Small

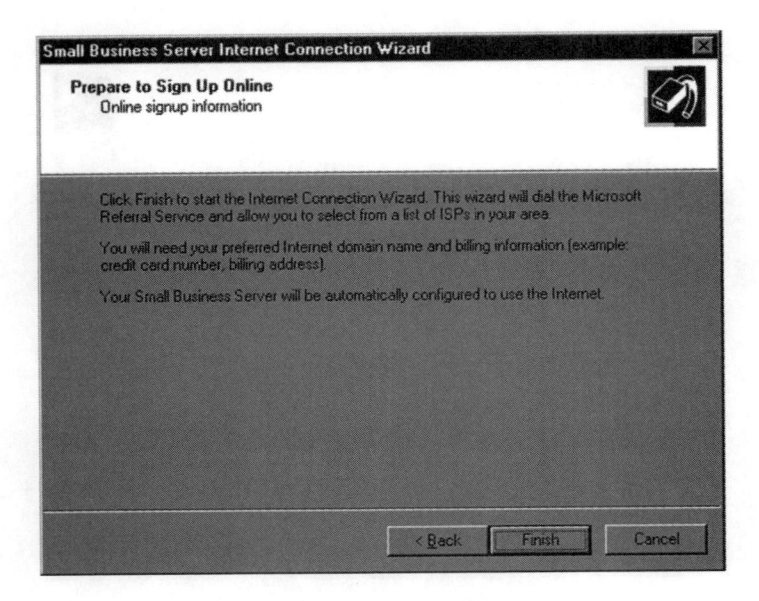

FIGURE 4 The Prepare to Sign Up Online page.

Business Server installs the DHCP server by default, and Windows NT (which is part of BackOffice Small Business Server) does not currently support using the DHCP server and the DHCP client on the same computer.

THE PREPARE TO SIGN UP ONLINE PAGE

This screen is a simple reminder to gather the information from your ISP to continue the installation. Clicking the Finish button will take you directly to the ICW, which automatically dials the Microsoft Referral Server using the configured modems and telephony properties. If more than one modem is configured for dial-out use in the Remote Access Service (RAS), you will be prompted to choose a modem with which to dial out.

THE SET UP MODEM CONNECTION TO ISP PAGE

When Modem or terminal adapter is selected on the Configure Hardware page, you will be presented with the Set Up Modem Connection to ISP page.

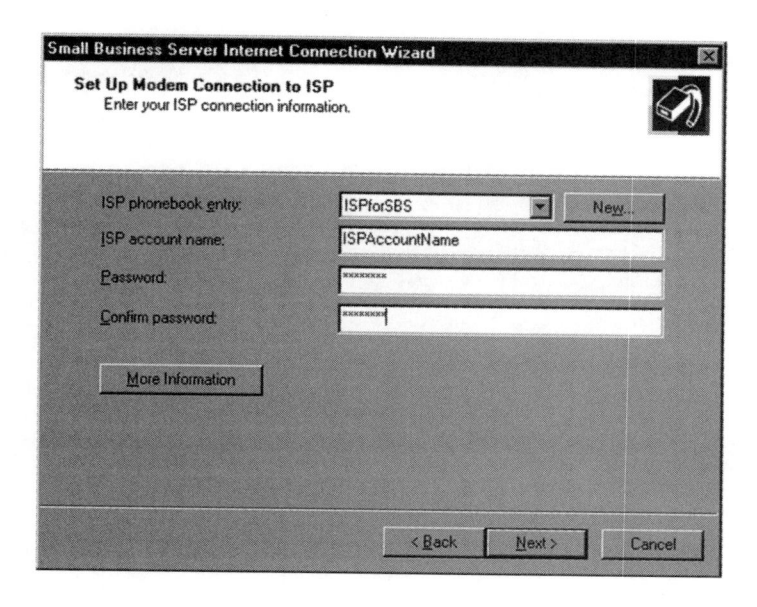

APPENDIX B: USING THE INTERNET CONNECTION WIZARD

FIGURE 5 The Set Up Modem Connection to ISP page.

If a phonebook entry has been previously created to connect to the ISP, you can select that entry. You will need to enter the credentials (account name and password) so that Proxy Server and the Exchange Server Internet Mail Service can connect when automatically dialing the ISP. In Internet Service Manager, these credentials are configured on the Credentials tab for Proxy Autodial for the Web Proxy or WinSock Proxy properties. These credentials are also set in Microsoft Exchange Administrator, under the Dial-up Connections tab of the Internet Mail Service.

Clicking the New button will launch the Dial-up Networking New Phonebook Entry Wizard to assist in the creation of a new phonebook entry (Figure 6).

THE SET UP ROUTER CONNECTION TO ISP PAGE

The Set Up Router Connection to ISP page gathers the information needed to correctly configure Small Business Server for use with a router on the network. The router address, which enables TCP/IP connectivity from the Small Business Server-based network to the Internet, is needed regardless of whether the router

FIGURE 6 The Set Up Router Connection to ISP page.

is connected directly to the local network or to the second network adapter configured under Small Business Server. These two situations are described in greater detail in the "Router Configuration" section of the introduction.

If you are going to configure Small Business Server to be the gateway to the Internet for the client machines, you will need to check the option My router is connected to the Small Business Server via a second network adapter. This option allows you to configure Small Business Server as a firewall.

If your router is not a full-time connection (for example, if it is a demand-dial ISDN router), select the My router is a dial-on-demand router option. When this option is selected, the Exchange Internet Mail Service will be configured to forward outgoing mail to the SMTP relay host specified later in the Configure SMTP Mail Delivery page.

THE NETWORK INTERFACE CARD CONFIGURATION PAGE

The Network Interface Card Configuration page is presented when Small Business Server is configured as the gateway for the local network. This page

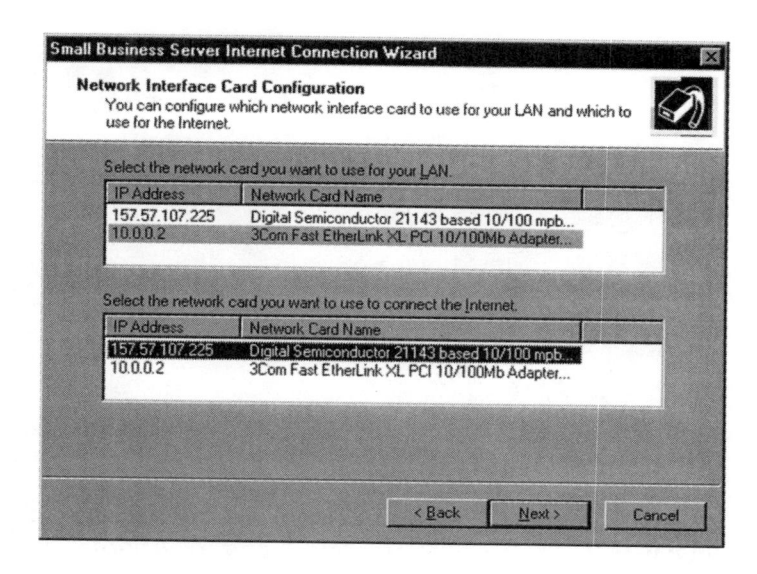

APPENDIX B: USING THE INTERNET CONNECTION WIZARD

FIGURE 7 The Network Interface Card Configuration page.

presents the network adapters in two windows. In the first window, you identify the internal network adapter that will be used for the local network. This internal adapter should be manually configured in the Local Address Table (LAT) in the Proxy Server configuration, using the Microsoft Management Console. In the second window, select the external network adapter that will be used to connect to the Internet. If you have more than one network adapter or have changed your IP address from the default address of 10.0.0.2, you will want to confirm the LAT configuration in Proxy Server (Figure 8).

THE SET UP SECOND NETWORK ADAPTER PAGE

Once the external network adapter is selected, the Set Up a Second Network Adapter page appears so that the adapter can be configured to work with a broadband device or router. Before the ICW is run, the external network card must be configured with a static IP address. (This is required because the alternative—receiving an address from the server's DHCP service—would require running the DHCP client on the DHCP server, which is not currently supported.) The default gateway and DNS servers should be con-

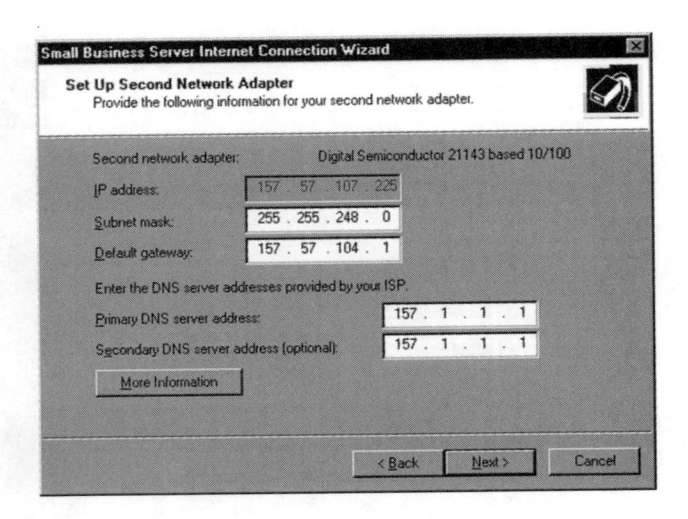

FIGURE 8 The Set Up Second Network Adapter page.

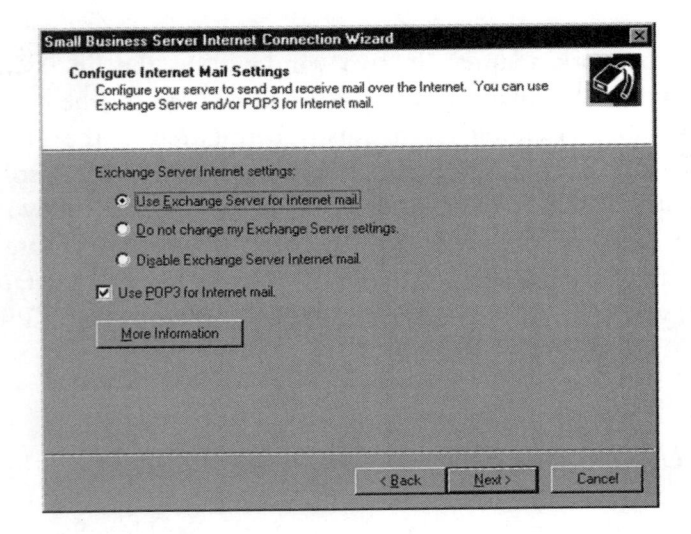

FIGURE 9 The Configure Internet Mail Settings page.

figured at this point if they were not configured through the Network area of Control Panel. These settings are required to complete Internet connectivity setup (Figure 9).

The Configure Internet Mail Settings Page

Once the networking pieces have been configured, the next step establish the e-mail configurations. The Configure Internet Mail Settings page configures Small Business Server to use Internet e-mail. If you are using SMTP e-mail through Exchange Server, select the Use Exchange Server for Internet mail option, which activates the Microsoft Exchange Internet Mail Service in Small Business Server. This is the option to choose, whether you host your own Internet domain and SMTP mail through a full-time connection or you dial in to an ISP that queues your mail until you connect. Selecting Disable Exchange Server Internet Mail will disable the Microsoft Exchange Server Internet Mail Service. This option disables only Internet mail through Exchange Server; it does not affect the ability to send mail to others on the local Small Business Server–based network. If you have already configured or customized your Exchange Server settings and do not wish to override these settings, select Do not change my Exchange Server settings.

If you will be retrieving e-mail from POP3 mailboxes at your ISP instead of directly through SMTP, select the Use POP3 for Internet mail option.

FIGURE 10 The Configure SMTP Mail Delivery page.

The Configure SMTP Mail Delivery Page

If in the Configure Internet Mail Settings page you select the option Use Exchange Server for Internet mail, the wizard presents the Configure SMTP Mail Delivery page. This page will allow you to configure the sending properties of the Microsoft Exchange Internet Mail Service. The choices available to you depend on the connection type:

- If your network is configured for a full-time connection (via router or broadband device), select the Use domain name system (DNS) for message delivery option.

- If your network is configured with a demand-dial router connection, select the Forward all mail to host option.

- If your network is configured with a dial-up connection, select the Forward all mail to host option.

If you do not have a full-time connection to the Internet, you should select Forward All Mail to Host, which covers dial-up connections, dial-on-demand routers, or any connection that does not maintain a full-time Internet connection. This option will set the Internet Mail Service to forward all messages to an SMTP relay host provided by your ISP. This relay host can be entered by its IP address or fully qualified domain name (for example, exchange1.isp.com or 192.168.16.1). If a dial-up connection is established through this wizard, the option to use DNS is unavailable.

If you have a full-time Internet connection using Exchange Server for your Internet mail, you should select Use domain name system (DNS) for message delivery. This option will use DNS queries to resolve any Internet e-mail addresses for delivery; thus the connection must be established before mail is sent (Figure 11).

The Receive Exchange Mail Page

When Exchange Server is used for SMTP mail, you must configure mail retrieval through the Receive Exchange Mail page. The correct settings can be hard to understand and must be negotiated with your ISP.

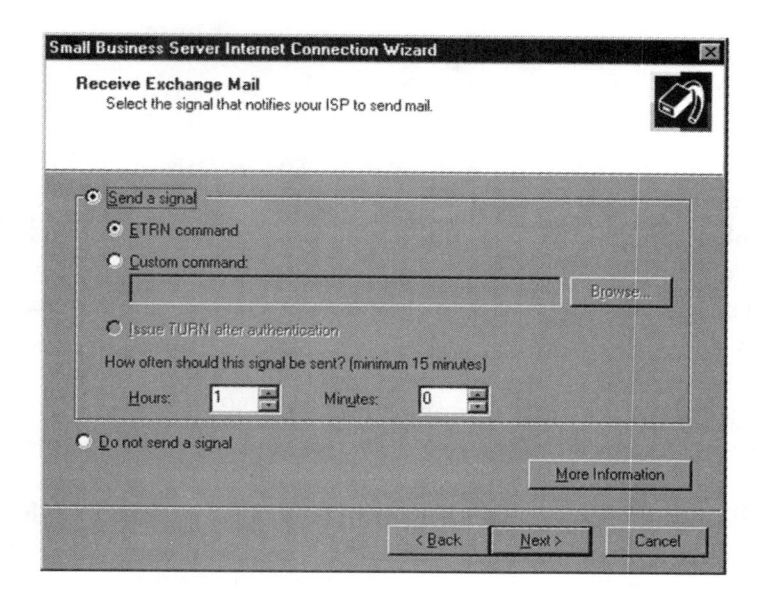

APPENDIX B: USING THE INTERNET CONNECTION WIZARD

FIGURE 11 The Receive Exchange Mail page.

When Small Business Server uses a full-time connection, it has a static IP address. The ISP's e-mail server uses that address to forward e-mail continuously as it arrives from other servers on the Internet. If your Small Business Server–based network has a full-time connection to the ISP, select the Do not send a signal option, and then click Next. This type of connection, however, does require that the MX and A records in the DNS database for your Internet domain point directly to your Exchange server. Because of the need for static A records, a full-time connection also requires a static IP address.

When Small Business Server has a dial-up connection, the ISP's e-mail server stores, or "queues," incoming messages, waiting for Small Business Server to create a connection and for Exchange Server to request that the e-mail server forward the stored messages. If your Small Business Server–based network uses a dial-up connection, select the Send a signal option, and then select the type of signal that the ISP's e-mail server expects Exchange Server to send. (You must negotiate the proper command with your ISP.) Also, select the frequency with which Exchange Server should attempt to send this signal. When you connect, Small Business Server issues a command to the ISP's

291

SMTP server to de-queue messages destined for your domain. Although the de-queue command is usually ETRN, which is supported by sendmail version 8.8.x or later, it could also be TURN after Authentication (sometimes known as ATRN) or a custom command given to you by the ISP. In any case, the de-queue command must be negotiated with your ISP to ensure compatibility. SBSETRN is one example of a custom command that de-queues mail. SBSETRN allows dial-up connections to receive a dynamic IP address when connecting to the ISP. Choosing Do not send a signal will not work in a dial-up scenario.

If ETRN is chosen for a dial-up connection, the ISP should configure an MX and an A record to point directly to the server running Small Business Server. An additional MX record would be added with a higher preference to point to the ISP's SMTP server. (The lower the preference number, the higher the priority.) This configuration allows the ISP to queue mail for Small Business Server until Small Business Server connects to the ISP, at which point it issues the ETRN command and de-queues mail. In this way, mail flows directly to Small Business Server upon connection. This configuration also provides the least amount of delay possible when receiving mail.

For more information about e-mail de-queuing methods, please see http://www.swinc.com/resource/exch_dq.htm. This table summarizes the configurations:

	Full-time connection	**Dial-up connection**	**Demand-dial router connection**
Fixed IP address	Do not send a signal	ETRN, or a custom command	ETRN, or a custom command
Dynamic IP address	Not supported	Custom command	Custom command

In terms of Exchange Server configuration, the demand-dial router connection is the same as a dial-up connection. Do not send a signal will not work with this scenario because the MX and A records for the Internet domain would point to Small Business Server, and incoming mail would never reach the Small Business Server–based network if the router was not connected. In this case, you must use the command given to you by the ISP. ETRN is the command the most likely to be used for demand-dial routers and static IP addresses. If Small Business Server does not have a static IP address, ETRN will not work, so the ISP must provide a custom command to de-queue e-mail.

From this page, you can set the frequency with which the signal is sent—that is, how much time elapses between attempted connections. This signal is also sent every time the connection is established with the ISP or every time you click Send and Receive Now on the Manage E-mail page of the Small Business Server Console. The default and recommended interval is one hour, and the minimum allowable interval is 15 minutes.

THE SET UP AUTHENTICATION PAGE

The Set Up Authentication page appears only when Exchange Server is used for Internet mail and the Issue TURN after authentication option is

FIGURE 12 The Set Up Authentication page.

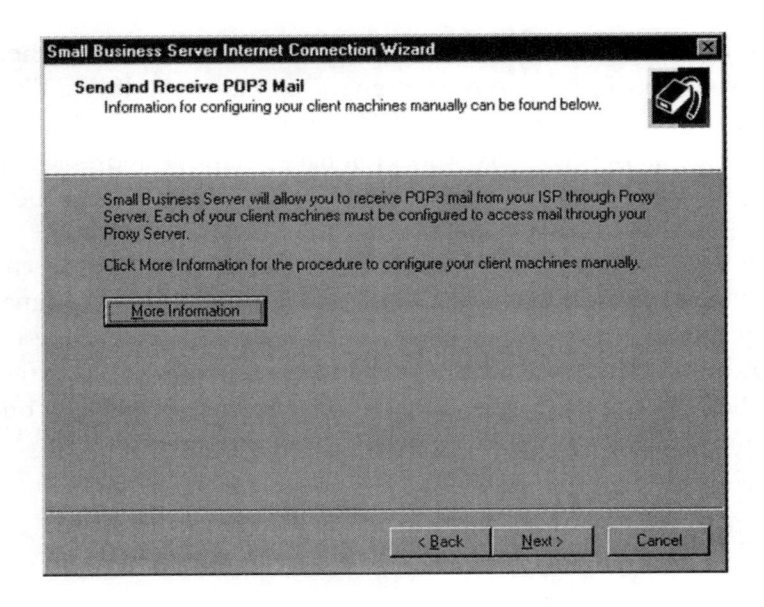

FIGURE 13 The Send and Receive POP3 Mail page.

selected on the Receive Exchange Mail page. TURN is a very secure dequeue signal that can be used for dial-up SMTP connections or for demand-dial router connections; do not attempt to use it for full-time router or broadband connections. Select the Issue TURN after authentication option only if your ISP supports this method and has requested that you use it, and you have a dial-up or demand-dial connection to the ISP. If this signal is used, the Use Secure Sockets Layer (SSL) option on the Set Up Authentication page should be selected; turning it off defeats the security benefits of TURN.

The TURN method of signaling is sometimes preferred over other options because it simplifies the communication between the ISP and Small Business Server. In particular, it allows message transfer without requiring the ISP's e-mail server to create a session with Small Business Server, so there is no need for the ISP's e-mail server to resolve the server name used by Small Business Server to an IP address. Communication between the ISP and the Small Business Server–based system over a dial-up connection is simpler when this name resolution isn't involved.

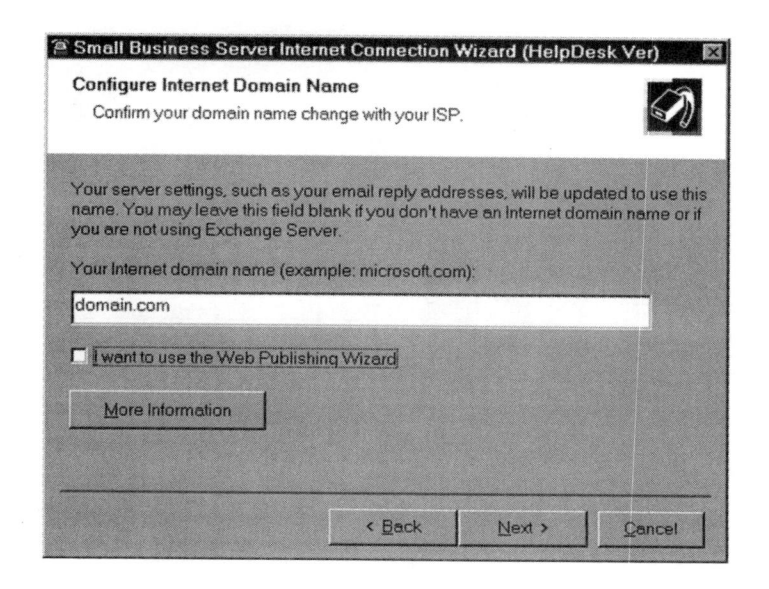

FIGURE 14 The Configure Internet Domain Name page.

The Send and Receive POP3 Mail Page

Client systems use the POP3 protocol to retrieve e-mail stored on a server. In the Small Business Server environment, Proxy Server manages the connection to the ISP, preventing outsiders from reaching the LAN and optionally preventing LAN users from reaching the Internet.

When you create user accounts for Small Business Server, you also configure the users' computers. If your users need access to POP3 e-mail accounts on the ISP's e-mail server, you need to configure their machines accordingly. Give those users' computers the WinSock Proxy client software so they can reach the Proxy Server, and add Internet Mail manually to each of those users' profiles in the Microsoft Outlook messaging and collaboration client.

The Configure Internet Domain Name Page

If Exchange Server is configured for SMTP mail, you must configure the Configure Internet Domain Name page. The domain name entered must be

FIGURE 15 The Configure Web Site Information page.

one provided by your ISP or one you have registered with the Domain Name Registry. This domain name will be configured as the reply address for all the recipients created in the Exchange Server Administrator. The domain name appears under the Site Addressing properties for the site in Exchange Server. If your connection to the Internet is a dial-up connection to the ISP, this is the domain name that will be configured for mail retrieval. The domain name is configured automatically in the Dialup Connections tab under the properties for the Microsoft Exchange Server Internet Mail Service in the Connections container for the site. Checking the "I want to use the Web Publishing Wizard" box displays a page for configuring your Web site information.

THE CONFIGURE WEB SITE INFORMATION PAGE

The Configure Web Site Information page gathers the information the system requires so that you can use the Web Publishing Wizard to create and update your Web site. In the Web page URL text box, enter the address on the Internet for your Web site. In the Web posting URL text box, enter the address that houses the data that makes up your Web site. If your ISP is currently hosting your Internet domain, and you would like to change your

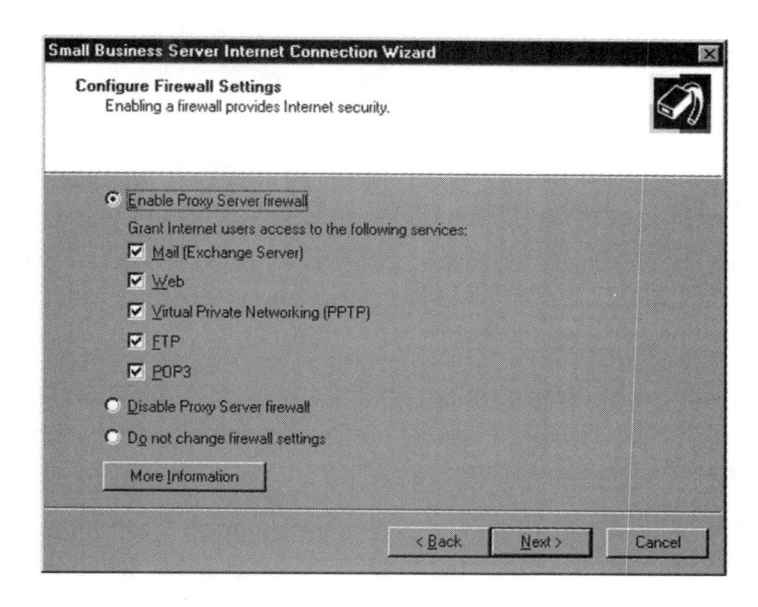

FIGURE 16 The Configure Firewall Settings page.

domain name, you must contact your ISP to request this change. If you are hosting your own Internet domain, your new domain name must be registered with the Domain Name Registry. In the Web posting account name and Password boxes, type this information, which is needed when you update your Web site. If your ISP does your Web hosting, this information should be provided by your ISP. In most cases, this account name and password are separate from your dial-up account name.

THE CONFIGURE FIREWALL SETTINGS PAGE

The final component of the Small Business Server Internet Connection Wizard is the Configure Firewall Settings page, which secures your local network by configuring Microsoft Proxy Server. If Enable Proxy Server firewall is selected, packet filtering will be enabled and local network clients will be able to access the Web through the Web Proxy service or the WinSock Proxy service, but users on the Internet will be blocked from accessing services on your Small Business Server–based network. If you like, the following services can be made available to Internet users by selecting the appropriate checkboxes on the Configure Firewall Settings page:

- Mail allows Small Business Server to listen on TCP port 25 (SMTP) to exchange mail with your ISP's SMTP server; that is, this option enables Small Business Server to exchange SMTP mail with your ISP. This checkbox must be selected if you are using Exchange Server for Internet mail.

- Web enables Small Business Server to listen to TCP port 80 (HTTP) and TCP port 443 (HTTPS), which allows Small Business Server to make Web pages accessible to users over the Internet.

- Virtual Private Networking (PPTP) opens the PPTP call and receive filters (TCP 1723) so that Internet clients can connect to the local network through a secure tunnel.

- FTP enables Small Business Server to listen on TCP ports 20 (FTP-data) and 21 (FTP), which allows Internet users to access the FTP service on Small Business Server.

- POP3 allows Small Business Server to listen to POP3 requests (TCP port 110) from the Internet.

The following packet filters are installed by default and are not configurable through the ICW:

DNSLookup

ICMP All Outbound

ICMP Ping Response

ICMP Ping Echo

ICMP Source Quench

ICMP Timeout

ICMP Unreachable

NetBIOS WINS Client

Note: The Point-to-Point Tunneling Protocol and FTP service are not installed by default on Small Business Server, but they may be installed if needed. These packet filter settings are accessible through the Security button on the Properties tab of the Web Proxy or WinSock Proxy service through the Microsoft Management Console of Microsoft Internet Information Server, the built-in Web server of Windows NT Server.

For more information regarding the features of Microsoft Proxy Server, refer to the Proxy Server online documentation.

REFERENCES

DEFINITION OF TERMS

The following terms are used in this document.

ADSL	asymmetric digital subscriber line
cable modem	A broadband connection device that enables high-speed Internet access
DHCP	Dynamic Host Configuration Protocol
DNS	Domain Name System, the method commonly used on the Internet for resolving host names to Internet address; a Domain Name Server is a server that uses the Domain Name System
DSL	digital subscriber line
DUN	Dial-up networking, the dial-out portion of RAS
ETRN	Extended TURN
firewall	A gateway machine dedicated to securing the network from outside attack, usually the Internet
FQDN	Fully Qualified Domain Name, the full DNS name of a system including its host name and domain name
FTP	File Transfer Protocol

HTTP	Hypertext Transport Protocol
ICW	Internet Connection Wizard
ISDN	Integrated Services Digital Network
ISP	Internet service provider
LAT	Local Address Table, a table consisting of the IP address ranges that define the internal network address space
NAT	Network Address Translation, which provides the ability to use private addresses internally by translating internal network addresses into a valid Internet address before transferring
NNTP	Network News Transfer Protocol, a protocol defined in RFC 977 for the distribution, inquiry, retrieval, and posting of Usenet news articles over the Internet
POP3	Post Office Protocol 3
PPTP	Point-to-Point Tunneling Protocol
RAS	Remote Access Service
router	A device that forwards packets between networks
SBSETRN	A custom version of ETRN designed by Microsoft
SMTP	Simple Mail Transfer Protocol
TCP	Transmission Control Protocol; TCP is the session level of the TCP/IP suite
TCP/IP	Transmission Control Protocol/Internet Protocol; TCP/IP usually refers to the protocol suite that includes TCP, UDP, and IP protocols
TURN with Authentication	An extension of the TURN command

URL   Uniform Resource Locator

VPN   Virtual private networking

DIAL-UP CONNECTION FLOW CHART

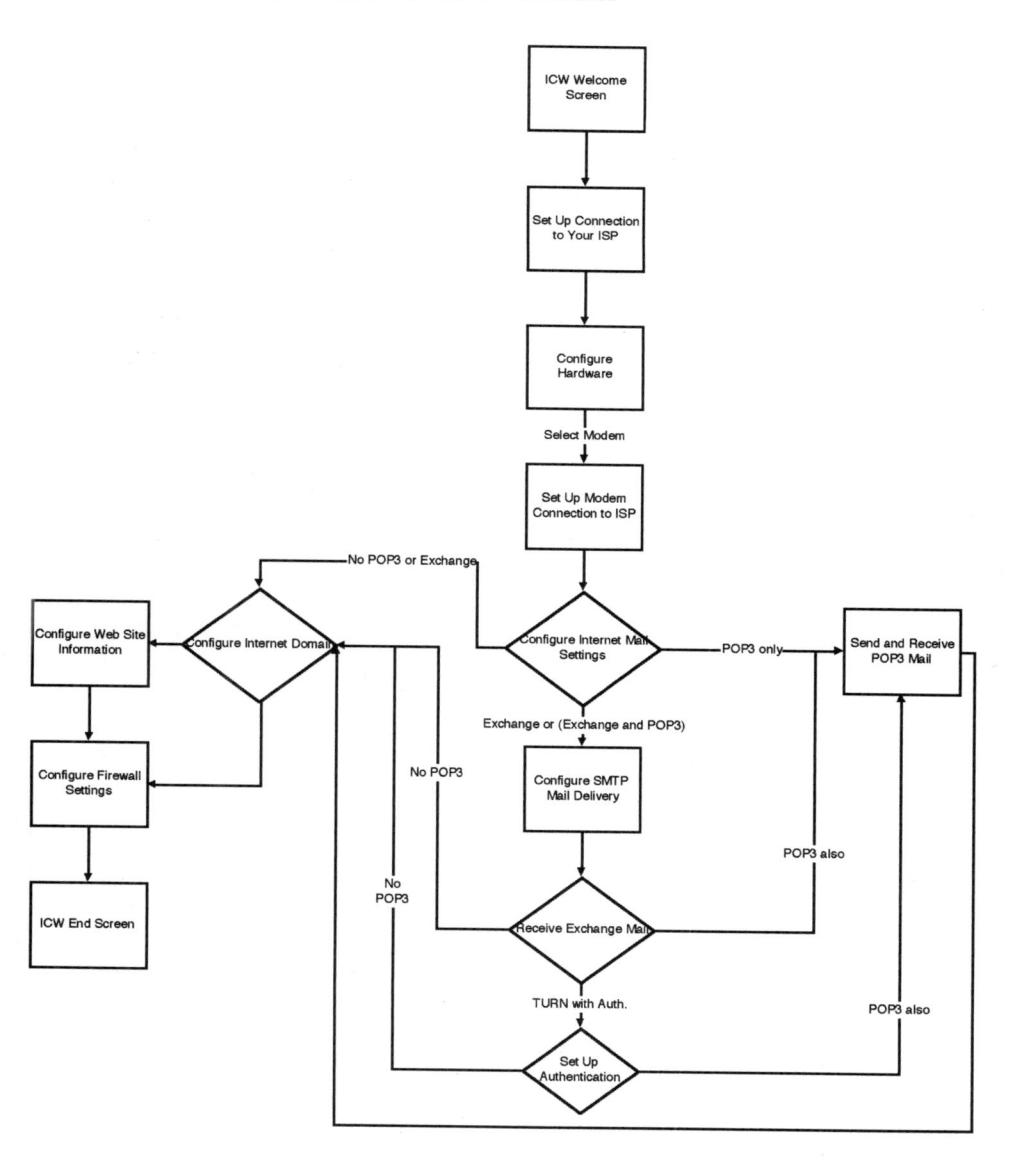

```
                              ┌──────────────┐
                              │ ICW Welcome  │
                              │    Screen    │
                              └──────┬───────┘
                                     ▼
                              ┌──────────────┐
                              │Set Up Conn.  │
                              │ to Your ISP  │
                              └──────┬───────┘
                                     ▼
                              ┌──────────────┐
                              │  Configure   │
                              │   Hardware   │
                              └──────┬───────┘
                                Select Modem
                                     ▼
                              ┌──────────────┐
                              │Set Up Modem  │
                              │Connection ISP│
                              └──────┬───────┘
                                     ▼
        ──No POP3 or Exchange──◇ Configure Internet Mail ◇──POP3 only──▶ Send/Receive POP3 Mail
                 │                     Settings
                 ▼                        │
        ◇ Configure Internet Domain   Exchange or (Exchange and POP3)
         │                                ▼
  Configure Web Site                ┌──────────────┐
   Information                      │Configure SMTP│
         │                          │ Mail Delivery│
         ▼                          └──────┬───────┘
  Configure Firewall                       ▼
     Settings                      ◇ Receive Exchange Mail ◇
         │                              TURN with Auth.
         ▼                                  ▼
   ICW End Screen                   ◇ Set Up Authentication ◇
```

Router Connection Flow Chart

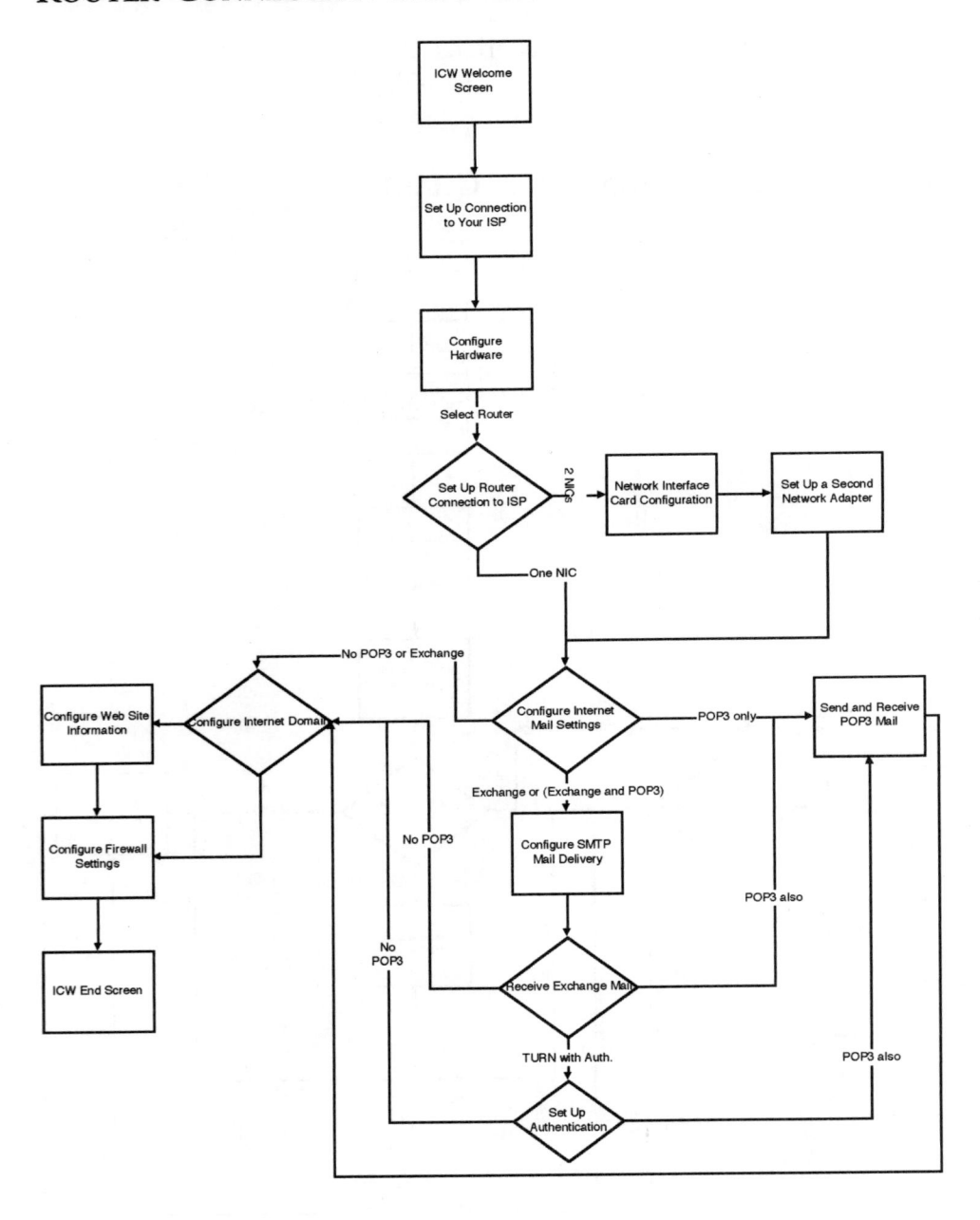

- ICW Welcome Screen
  - ↓
- Set Up Connection to Your ISP
  - ↓
- Configure Hardware
  - ↓ Select Router
- Set Up Router Connection to ISP
  - 2 NICs → Network Interface Card Configuration → Set Up a Second Network Adapter
  - One NIC ↓
- Configure Internet Mail Settings
  - POP3 only → Send and Receive POP3 Mail
  - No POP3 or Exchange → Configure Internet Domain
  - Exchange or (Exchange and POP3) ↓
- Configure SMTP Mail Delivery
  - ↓
- Receive Exchange Mail
  - No POP3 → Configure Internet Domain
  - POP3 also → Send and Receive POP3 Mail
  - TURN with Auth. ↓
- Set Up Authentication
  - POP3 also → Send and Receive POP3 Mail
  - → Configure Internet Domain

- Configure Internet Domain
  - → Configure Web Site Information
  - ↓
- Configure Firewall Settings
  - ↓
- ICW End Screen

APPENDIX B: USING THE INTERNET CONNECTION WIZARD

FULL-TIME/BROADBAND CONNECTION FLOW CHART

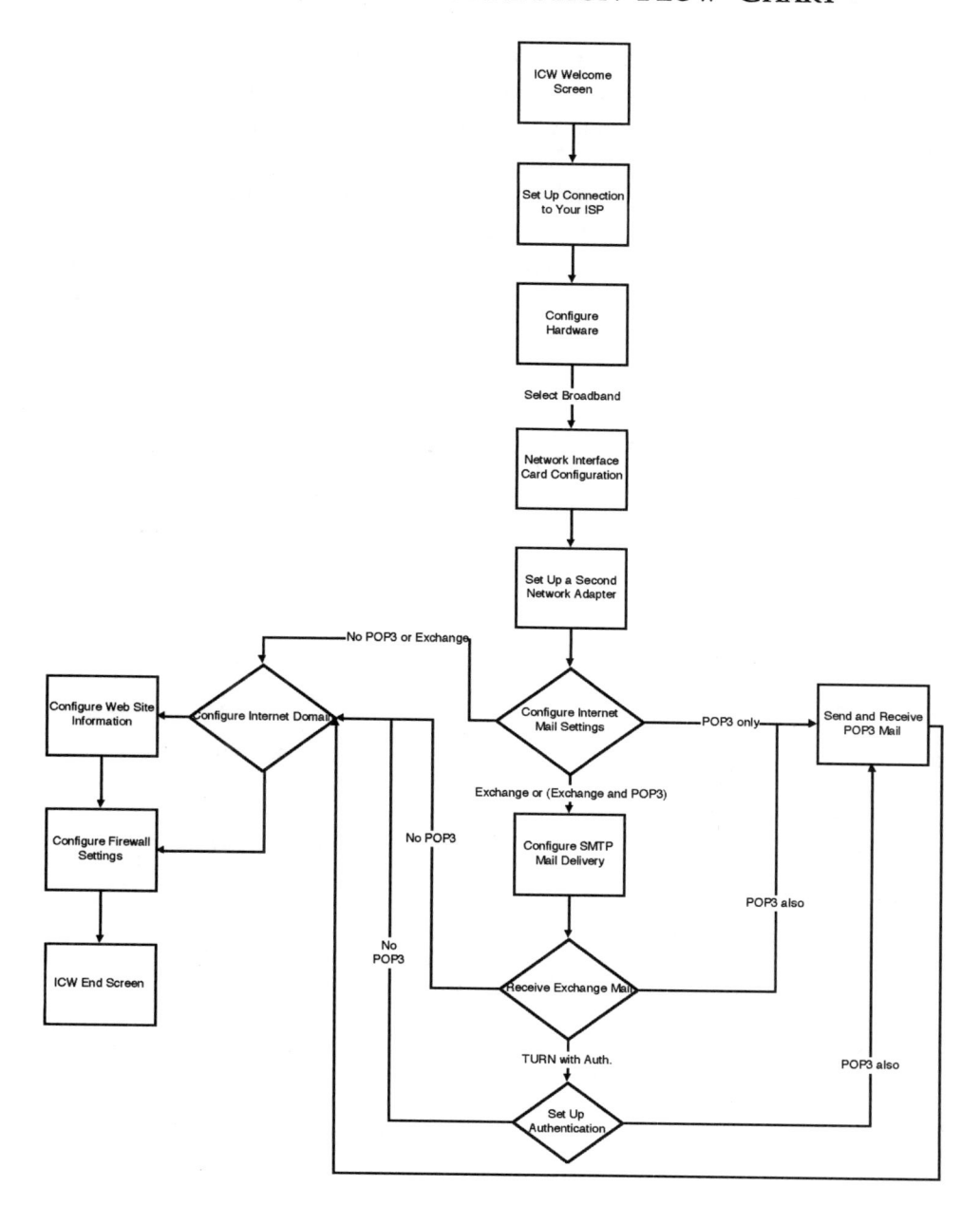

```
                            ICW Welcome
                               Screen
                                  |
                                  v
                          Set Up Connection
                             to Your ISP
                                  |
                                  v
                             Configure
                             Hardware
                                  |
                         Select Broadband
                                  |
                                  v
                          Network Interface
                         Card Configuration
                                  |
                                  v
                          Set Up a Second
                          Network Adapter
                                  |
          No POP3 or Exchange─────┤
                                  v
  Configure Web Site   Configure Internet   Configure Internet                Send and Receive
    Information    <──    Domain        <── Mail Settings    ──POP3 only──>    POP3 Mail
        |                                      |
        v                   No POP3    Exchange or (Exchange and POP3)
  Configure Firewall          |                 |
     Settings                 |                 v
        |                     |          Configure SMTP
        |                     |           Mail Delivery              POP3 also
        v                     |                 |
   ICW End Screen          No POP3              v
                              |          Receive Exchange Mail
                              |                 |
                              |           TURN with Auth.
                              |                 |
                              |                 v
                              └─────────  Set Up
                                        Authentication ────POP3 also──>
```

303

SBSETRN Definition and Parameters

If you are using a dial-up connection and receiving your IP address dynamically, you must use a custom command to de-queue your mail. One example of a custom command that handles dynamic IP addressing is Sbsetrn.exe. Sbsetrn.exe is a special form of ETRN designed and implemented by Microsoft. This command allows for the delivery of mail to SMTP hosts with dynamically allocated IP addresses. Sbsetrn.exe is located in the %systemdrive%\SmallBusiness directory. Because your server is allocated an IP address dynamically, Sbsetrn.exe creates a WINS entry to identify your server to your ISP's mail server and then routes your mail appropriately. The parameters for Sbsetrn.exe are read from the registry unless otherwise specified. When scheduling the task, the parameters can be included with the function call or the parameters will be read from the registry.

Usage: sbsetrn [-n <name>] [-d <domain>] [-s <SMTP server>] [-l]

Parameters:

Flag	Required	Description	Corresponding registry key
n	Yes	Server NetBIOS name	HKLM\Software\Microsoft\Small Business\Internet\MS_IMS\Mail_DeliveryKey
d	No	Domain name	HKLM\Software\Microsoft\Small Business\Internet\MS_IMS\Mail_Domain_Name or HKLM\Software\Microsoft\Small Business\Internet\Domain Registration\Domain Name (if it exists)
s	No	ISP's SMTP server name	HKLM\Software\Microsoft\Small Business\Internet\MS_IMS\Mail_Route_Host_Name
w	Yes	IP address or name of WINS server	HKLM\Software\Microsoft\Small Business\Internet\MS_IMS\WINS_Address and HKLM\Software\Microsoft\Small Business\Internet\MS_IMS\WINS_Alt_Address
l	No	Local echo	
e		Delay	

Full-time/Broadband Connection Information Form

The following form outlines the information needed before using the Small Business Server Internet Connection Wizard.

Information about your second network adapter:

IP address: _____

Subnet mask: _____

Default gateway: _____

Primary DNS server address (provided by your ISP): _____

Secondary DNS server address (optional): _____

What type of mail do you receive?

    ___ Exchange Server (SMTP)   ___ POP3

If you are using Exchange Server, what service do you use to deliver your mail? _____

    ___ Forward all messages to host

Host name or IP address of your ISP's mail (SMTP) server (example: exchange.microsoft.com or 170.10.10.10) _____

DNS

Your Internet domain name (example: microsoft.com): _____

Your Web page URL (example: www.microsoft.com): _____

Your Web posting URL (if different): _____

Web posting account name: _____

Password: _____

CONNECTION WITH A MODEM INFORMATION FORM

The following form outlines the information needed before using the Small Business Server Internet Connection Wizard.

ISP phonebook entry: _____

ISP account name: _____

Password: _____

Note: If you are entering a new phonebook entry, you must contact your ISP for the configuration details to help ensure a successful connection.

What type of mail do you receive? (please choose one type)

___ Exchange Server (SMTP)　　___ POP3

If you are using Exchange Server (SMTP): _____

Host name or IP address of your ISP's mail (SMTP) server (example: exchange.microsoft.com or 170.10.10.10)

_____

How do you signal your ISP to send mail to you?

___ ETRN

___ Custom command file location: _____

___ TURN with authentication

___ No signal

Note: If you are using TURN with authentication, you will need to know your authentication name and password. This information is provided by your ISP.

How often should this signal be sent? _____

Your Internet domain name
(example: microsoft.com): _____

Your Web page URL
(example: www.microsoft.com): _____

Your Web posting URL (if different): _____

Web posting account name: _____

Password: _____

CONNECTION WITH A ROUTER INFORMATION FORM

The following form outlines the information needed before using the Small Business Server Internet Connection Wizard.

IP address for your router (example: 10.0.0.1): ____.____.____.____

Primary DNS server address
(provided by your ISP): ____.____.____.____
Secondary DNS server address (optional): ____.____.____.____

Do you have a dial-on-demand router (that is, a nonpermanent connection to the Internet)?

    ___ Yes   ___ No

Is your router connected to your server through a second network adapter?

    ___ Yes   ___ No

What type of mail do you receive? (please choose one type)

    ___ Use Exchange Server (SMTP)   ___ POP3

If you are using Exchange Server (SMTP), what service do you use to deliver your mail? _____

Forward all messages to host

Host name or IP address of your ISP's mail (SMTP) server
(example: exchange.microsoft.com or 170.10.10.10)

_____

DNS

If you are using Exchange Server, a dial-on-demand router and your messages are forwarded to your ISP's mail server:

How do you signal your ISP to send mail to you?_____

ETRN _____

Custom command file location: _____

No signal _____

How often should this signal be sent? _____

Your Internet domain name: _____

Your Web page URL: _____

Your Web posting URL (if different): _____

Web posting account name: _____

Password: _____

Glossary

10BaseT—Also known as the IEEE 802.3, this is the original 10- Mbps Ethernet standard.

100BaseTX—This is the networking standard for 100-Mbps Fast Ethernet that uses Category-5 cables.

Access point—Linksys' wireless-based device for connecting roaming wireless PC cards directly to the Internet. The access point is a device that provides the benefits and mobility of roaming from a stationary Internet connection.

Asymmetric digital subscriber line (ADSL)—A DSL technology providing asymmetrical bandwidth over a single wire pair. The downstream bandwidth going from the network to the subscriber is typically greater than the upstream bandwidth going from the subscriber to the network.

Asynchronous transfer mode (ATM)—Under ATM, multiple traffic types (such as voice, video, or data) are conveyed in fixed-length cells (rather than the random-length packets moved by technologies such as Ethernet and Fiber Distributed Data Interface [FDDI]). This enables very high speeds, making ATM popular for demanding network backbones. With the networking equipment that has recently become available, ATM will also support WAN transmissions. This feature makes ATM valuable for large, dispersed organizations.

Backbone—The part of a network that acts as the primary path for traffic moving between, rather than within, networks.

Bandwidth—The data-carrying capacity of a network connection, used as an indication of speed. For example, an Ethernet link is capable of moving data at a rate of 10 Mbps. A Fast Ethernet link can move data at a rate of 100 Mbps—10 times more bandwidth.

Bridge—A device that passes packets between multiple network segments using the same communications protocol. If a packet is destined for a user within the sender's own network segment, the bridge keeps the packet local. If the packet is bound for another segment, the bridge passes the packet onto the network backbone.

Cable modem—A class of modem that is used for connecting to a cable TV network, which in turnconnects directly to the Internet. Cable-modem-based connections to the Internet are typically much faster than dial-up modems, yet they have the issue of security, as a cable-based network is comparable to a closed network.

Client—A networked PC or terminal that shares services with other PCs. These services are stored on or administered by a server.

Digital subscriber line—A digital phone service that provides for voice, video, and digital data over existing phone systems at higher speeds than are available in typical dial-up Internet sessions. This technology delivers high bandwidth over conventional copper wiring at limited distances. There are four types of DSL: asymmetric digital subscriber line (ADSL), high-speed digital subscriber line (HDSL), symmetric digital subsriber line (SDSL), and very high-speed digital subscriber line (VDSL). All are provisioned via modem pairs, with one modem located at a central office and the other at the customer site. Because most DSL technologies do not use the whole bandwidth of the twisted pair, there is room remaining for a voice channel.

DSL modem—A modem that connects a PC to a DSL network, which in turn connects to the Internet.

Ethernet—A popular LAN technology that uses CSMA/CD (collision detection) to move packets between workstations and runs over a variety of cable types at 10 Mbps. Also called 10Base-T.

Extranet—A network that gives external users (such as suppliers, independent sales agents, and dealers) access to company documents such as price lists, inventory reports, shipping schedules, and more.

Fast Ethernet—Uses the same transmission method as 10-Mbps Ethernet (collision detection) but operates at 100 Mbps—10 times faster. Fast Ethernet provides a smooth upgrade path for increasing performance in congested Ethernet networks because it uses the same cabling, applications, and network management tools. Variations include 100Base-FX, 100Base-T4, and 100Base-TX.

Fiber Distributed Data Interface (FDDI)— A LAN technology based on a 100-Mbps token-passing network running over fiber-optic cable. Usually reserved for network backbones in larger organizations.

File Transfer Protocol (FTP)—A part of the chief Internet protocol stack or group (TCP/IP), used for transferring files from Internet servers to your computer.

Frame relay—WAN service that provides switched (on-and-off) connections between distant locations.

Gigabit Ethernet—The latest version of Ethernet. It offers 1,000 Mbps or 1 Gbps of raw bandwidth. It is 100 times faster than the original Ethernet, yet it is compatible with existing Ethernets because it uses the same CSMA/CD and Media Access Control (MAC) protocols. Gigabit Ethernet competes most directly with ATM and is forcing out FDDI and Token Ring.

Home Phoneline Networking Alliance (HomePNA)—An organization that works to ensure that all products sold into the home networking marketplace adopt a single, unified phone-line networking standard. This is specifically done to bring a unified set of interoperable home networking solutions to the marketplace. Linksys is a member of the HomePNA association.

Hub—A networking device that enables attached devices to receive data streams that are transmitted over a network. A hub also makes it possible for devices to share the network bandwidth available on a network. A hub interconnects clients and servers, repeating (or amplifying) the signals between them. Hubs act as wiring *concentrators* in networks based on star topologies (rather than bus topologies, in which computers are daisy-chained together).

Hypertext Markup Language (HTML)—A simple document-formatting language used for preparing documents to be viewed by a tool such as a WorldWideWeb browser.

Hypertext Transfer Protocol (HTTP)—A protocol that governs transmission of formatted documents over the Internet.

Integrated Services Digital Network (ISDN)—A communication protocol offered by telephone companies that permits high-speed connections between computers and the network in dispersed locations.

Internet—A massive global network, interconnecting tens of thousands of computers and networks worldwide and accessible from any computer with a modem or router connection and the appropriate software.

Internet Service Provider (ISP)—Typically a company or organization that provides Internet access for individuals and companies.

Intranet—An internal network that takes advantage of some of the same tools popularized on the Internet (browsers for viewing material, HTML for preparing company directories or announcements, and so on).

IP telephony—IP telephony combines different types of communications—such as data, voice, and video—over a single-packet cell-based infrastructure. IP telephony extends the value of the network with these nontraditional applications. By combining different types of traffic on a single network connection, small- and medium-sized businesses and small branch offices can dramatically reduce the cost of their voice and data networks.

ISDN digital subscriber line (IDSL)—A DSL technology that is basically a naming convention for an ISDN Basic Rate Interface (BRI), both B channels and the D channels permanently bonded for 144 Kbps over a single wire pair. IDSL uses 2B1Q line coding.

GLOSSARY

Local Area Network (LAN)—A series of PCs that have been joined together via cabling so that resources can be shared, including file and print services; typically, a network or group of network segments confined to one building or a campus (compare to WAN). LANs are increasingly being found in homes, where sharing of Internet access is one of the most dominant uses of this networking approach.

Megabits per second (Mbps)—Defines the speed at which data is traveling, which is measured in millions of bits per second. This is a measure of the performance of a device.

Modem—Device that enables a computer to connect to other computers and networks using ordinary phone lines. Modems modulate the digital signals of the computer into analog signals for transmission, and then demodulates those analog signals back into the digital signals that the computer on the other end can understand.

Network—Typically a collection of devices, including PCs, printers, and storage devices, that are connected together for the purpose of sharing information and resources.

Network Interface Card (NIC)—A device that provides for connecting a PC to a network. NIC cards, also called network adapters, provide the essential link between a device and the network. NICs are also found in many peripherals, including storage subsystems and printers.

Packet—A block of data with a header attached that can indicate what the packet contains and where it is headed. Think of a packet as an envelope filled with data, with the header acting as an address on the envelope.

Remote access server—Device that handles multiple incoming calls from remote users who need access to central network resources. A remote access server can enable users to dial into a network using a single phone number. The server then finds an open channel and makes a connection without returning a busy signal.

Router—Device that moves data between different network segments and can look into a packet header to determine the best path for the packet to travel. Routers can connect network segments that use different protocols. They also enable all users in a network to share a single connection to the Internet or a WAN.

Server—A computer or even a software program that provides services to clients. These services might include file storage (file server), programs (application server), printer sharing (print server), and fax (fax server) or modem sharing (modem server). Also see *Client*.

Switch—A device that improves network performance by segmenting the network and reducing competition for bandwidth. When a switch port receives data packets, it forwards those packets only to the appropriate port for the intended recipient. This further reduces competition for bandwidth between the clients, servers, or workgroups connected to each switch port.

Token Ring—LAN technology in which packets are conveyed between network end stations by a token moving continuously around a closed ring between all the stations. Runs at 4 or 16 Mbps.

Virtual private network (VPN)—Enables IP traffic to travel securely over a public TCP/IP network by encrypting all traffic from one network to another. A VPN uses *tunneling* to encrypt all information at the IP level.

Wide Area Network (WAN)—A public or private network that provides coverage of a broad (wide) geographic area. WANs are typically used for connecting several metro areas as part of a larger network. Universities and larger corporations typically use WANs to connect their geographically dispersed locations.

Index

A

Access Control List 246
access point 140, 237
access point transceiver 140
Add/Remove Programs 202
address 103
address, broadcast 105
address, host 103, 106, 110
Address Resolution Protocol 112
address, subnet 110
ADSL 195, 299
American Registry for Internet
  Numbers 110
antenna, directional 30
AppleTalk 133, 170
Appletalk 87
application layer
  57, 164, 171, 172
APPN Network Node 90
ARIN 110
ARP 112
ARPA 112
ARPANET 48, 205
Assigned Services RFC 62
asymmetric digital subscriber line
  299

Asymmetric DSL 195
Asynchronous Transfer Mode 95
ATM 95
Attachment Unit Interface 130
attenuation 124
AUI 130
autonomous system 175

B

B-node 212
bandwidth 39, 122, 124
barker sequence 240
baseband 123
Basic Service Set 237
beacon 246
BGP 62, 170
Binary Phase Shift Keying 240
Bluetooth 18, 21, 267
BNC 129
BOOT P 75, 206
Border Gateway Protocol 62, 170
bounded media 119
boundless media 120
BPDU 174
BPSK 240

bridge 90
bridge protocol data unit 174
bridging, wired 24
bridging, wireless 24
British Naval Connector 129
broadband 2, 3, 123
broadband modem 195
broadcast address 105
broadcast name resolution mode 212
BSS 237
byte 103

C

cable, coax 85, 89
cable, coaxial 39, 42, 125, 126, 129, 263
cable, fiber-optic 85, 121, 124, 125, 136, 137, 145
cable media 120, 145
cable modem 39, 299
cable, RS-422 98
cable, STP 133
cable tester 151, 155
cable, transceiver 130
cable, twisted-pair 125, 131
Cable, Unshielded Twisted-Pair 133
cable, UTP 39, 137, 151
cable-based network 142
cables, RS-232 98
cabling, daisy-chain 128
card, Ethernet 85
Carrier Sense Multiple Access with Collision Avoid 17, 242

Carrier Sense Multiple Access with Collision Detect 14, 241
CCK 240
CDPD 144
CeBus 268
cell 174
cellular digital packet data 144
channel reuse 244
CLEC 265
CMIP 172
coax cable 85, 89
Coaxial cable 125
coaxial cable 39, 42, 125, 126, 128, 129, 263
command syntax 147
Common Management Information Protocol 172
competitive local exchange carrier 265
Complementary Code Keying 240
computer-based network 189
connector, RJ-45 151
Consumer Electronic Bus 268
Continuous Aware Mode 245
control information 165
Control Panel 202
converter 98, 100
CRC 55, 185
cross talk 125
CSMA/CA 17, 242
CSMA/CD 14, 88, 242
cyclic redundancy check 55, 185

D

D-Link Systems 6
daisy-chain cabling 128

daisy-chain wiring 40
Data Over Cable Systems Interface Specifications 267
data packet 53
data-link layer 55, 162, 164, 166, 241
datagram 53, 62, 173
DECNET 88, 93
dedicated gateway 191
dedicated home gateway 192
Device Server 99
DHCP 48, 50, 58, 73, 75, 199, 203, 277, 299
DHCP server 260
dial-up networking 276, 299
dial-up modem 3
digital subscriber line 39, 188, 195, 267, 299
direct sequence 25
direct-sequence spread spectrum 14
directed name resolution mode 213
directional antenna 30
Distributed Coordination Function 242
DNS 48, 80, 81, 205, 299
DNS server 279
domain 175
Domain Name Registry 297
Domain Name Server 279
Domain Name System 299
Domain Naming Services 48, 80
DOS 86
DSL 39, 188, 267, 299
DSL modem 194, 195, 262
DSSS 14, 25

DSSS modulation 14
Dynamic Host Configuration Protocol 48, 50, 277, 299
dynamic IP addressing 206
dynamic rate shifting 241

E

e-commerce 256, 260
EGP 61
electromagnetic waveform 120
electromagnetic interference 124
EM spectrum 120
EM waveform 120
EMI 124, 126, 128, 131
encapsulation 166
end system 175
error checking 170
error recovery 170
ESS 237
ESSID 246, 252
Ethernet 4, 8, 10, 20, 76, 85, 88, 94, 152, 236, 243
Ethernet adapter 92
Ethernet card 85
Ethernet II 86
Ethernet LAN 23, 91, 123
Ethernet network 4
ETRN 299
Exchange Server 279
Extended Service Set 237
Extended Service Set Identification 246
Extended TURN 299
Exterior Gateway Protocol 61

F

FDDI 94
FHSS 14, 16, 19, 25
Fiber Distributed Data Interface 94
Fiber-optic cable 125
fiber-optic cable 85, 91, 121, 124, 136, 137, 145
fiber-optic network 136
File Transfer, Access, and Management 172
File Transfer Protocol 51, 299
firewall 299
flow control 170
focused transmission 120
FQDN 205, 299
frame 55, 173
frequency division multiplexing 10
frequency hopping 25
frequency-hopping spread spectrum 14
FTAM 172
FTP 51, 65, 66, 152, 172, 197, 299
fully qualified domain name 205, 299

G

gateway 189
gateway, dedicated 191
gateway, dedicated home 192
gateway-based network 189
Gateway-to-Gateway Protocol 61
GGP 61
GIF 172

Graphics Interchange Format 172

H

H-node 214
hidden node 242
hierarchical address 181
home agent 250
home gateway 255, 262
home gateway services 262
Home RF 227
home run 40
HomeCast Open Protocol 19
HomePlug Powerline Alliance 268
HomePNA 8, 9, 11, 12, 32
HomeRF 10, 16, 18, 32, 267
HOP 19
host address 103, 106, 110
HOSTS file 82
HTTP 300
hub 100
Hypertext Transport Protocol 300

I

IAN 8
IANA 184
IBSS 238
ICMP 61, 113, 115, 150
IEEE 802 236
IEEE 802.11 15, 16, 227
IEEE 802.11b 36
IEEE 802.15 22
IEEE 802.2 169
IEEE 802.3 20
IETF 87
IIS 66

impedance 126
Independent Basic Service Set 238
Industry Standard Architecture 20
Information Appliance Network 8
information exchange process 166
Infra-red Data Association 44
Infrared 26, 43
infrared 120, 263
infrared transmission 121
Integrated Services Digital
    Network 300
Intel 6
Interfrequency Switching Mode
    17
intermediate system 175
Internet 2, 3, 10, 47, 48, 58,
    82, 86, 87, 88, 98,
    110, 170, 189
Internet Assigned Numbers
    Authority 184
Internet Connection Wizard 280
Internet Control and Messaging
    Protocol 61, 115
Internet Engineering Task Force
    87
Internet Explorer 65
Internet Home Alliance 255, 270
Internet Information Server 66
Internet Network Information
    Center 58
Internet service provider 300
Internet Webtone 255
internetwork 159
InterNIC 58, 60
intranet 98
IP 93

IP address 50, 58, 60, 76,
    77, 148, 211, 279
IP header 112
IP packet 112
IP routing 61
IP stack 111
IPX 93
IPX.COM 86
IPX/SPX 55
IR 26
IR communication 142
IR technology 43
IrDA 44
ISA 20
ISDN 300
ISDN terminal adapter 276
ISM 17
ISP 50, 76, 265, 300

J

jack, RJ-45 13
Java Embedded Server software
    255, 270
Java Virtual machine 48
Joint Photographic Experts Group
    172
JPEG 172
JVM 48

K

Kingston 6

L

LAN 1, 16, 155, 168

LAN card 6
LAN, Ethernet 23
LAN Manager 223
LAN Manager HOSTS file 82
LAN, wireless 16, 20, 22, 24, 30
LAT 300
layer, application 57, 164, 171
layer, data-link 55, 162, 164, 166
layer, network 55, 56, 166
layer, peer 164
layer, physical 54, 163, 164, 168
layer, presentation 57, 164, 171
layer, session 56, 171
layer, transport 56
LCD display 151
LCD screen 151
LED 137
light-emitting diodes 137
line-of-sight transmission 120
Linksys 6
LLC 169
LMHOSTS 211, 226
LMOSTS 82
load balancing 243
Local Address Table 287, 300
Local Area Connection 209
local network 98
locally administered address 92
Logical Link Control 169
long-range wireless bridge 143
loop wiring 40
lower layer 162

M

m-commerce 227, 228, 229, 232
M-node 213
MAC 169
management information system 236
media access controller protocol 14
media, bounded 119
media, boundless 120
media, cable 120, 145
media, transmission 120
message 174
microcell 26
Microsoft Outlook 295
Microsoft Proxy Server 276
Microwave transmission 120
modem 3, 98, 195
modem, broadband 195
modem, cable 39, 299
modem, dial-up 3
modem, DSL 194, 195, 262
Motion Picture Experts Group 171
MPEG 171
multiplexing 58, 62, 124, 170, 186
My Network Places 203, 208

N

N-connector 129
Name Binding Protocol 170
narrowband radio 143
NAT 111
NBP 170
NDIS 86
NetBEUI 54, 74
NetBIOS 82, 86, 92, 201, 211, 223
NETGEAR 6
NetWare 49, 55, 63, 92, 201, 236
network access provider 265
network address 181
Network Address Translation 111, 300
network, cable-based 142
Network Connections 208
Network Connections window 203
Network Device Interface Specification 86
network dialog box 202
network, Ethernet 4
Network Interface Card 8
network layer 55, 56, 166
Network News Transfer Protocol 197, 300
network, peer-to-peer 3
network, power-line 6
network topology 169
network-layer address 181
NIC 8
NIC card 76
NNTP 197, 300

O

octet 103, 107
ODI 86
OEM 8
OFDM 15
Open Services Gateway initiative 255, 268
Open Shortest Path First 170
Open Systems Interconnect Model 51
Original Equipment Manufacturer 8
orthogonal frequency division multiplexing 15
OSGi 255, 268, 271
OSI application 172
OSI layer 165
OSI Model 51, 52, 53, 57, 111, 159, 161, 172, 236
OSI protocol suite 170
OSI transport protocol 170
OSPF 170
Output Device Interface 86

P

P-node 213
P2P 3
packet 173
parallel port 8
PCI 6
PCI card 12, 36
PDU 174
peer layer 164
Peer Web Services 65
peer-to-peer network 3, 27

Perform Incremental Backup 221
peripheral 3
Peripheral Component Interconnect 6
peripheral device 43
Phone-line home networking 9
physical layer 54, 163, 164, 168, 238
Point Coordination Function 244
point-to-point communication 138
point-to-point connection 138
Point-to-Point Tunneling Protocol 298, 300
polyvinyl chloride 131
POP3 279, 295, 300
Post Office Protocol 3 300
POTS 10
Power Packet Home Networking Technology 9
Power Save Polling Mode 245
Power Saving Protocol 251
power-line modem chip 6
power-line network 6
power-line technology 6
PPTP 298, 300
pread-spectrum radio 25
presentation layer 57, 164, 171
propagation delay 145
Protec 6
protocol data unit 174
protocol stack 148
PSTN 189
public switched telephone network 189
PVC cabling 131

Q

QoS 10, 20, 100
QPSK 240
quad wiring 38
Quadrature Phase Shift Keying 240
quality of service 10
QuickTime 171

R

RadioLAN 20
RAS 74, 300
Reduced Instruction Set Computing 66
Remote Access Service 300
Remote Access Services 74, 276
Request to Send/Clear to Send 242
Requests for Comments 51
resistor 128
RFCs 51
RG-59 coax 43
RG-6 coax 42
ring topology 169
RIP 153, 170
RISC 66
RJ-45 connector 151
RJ-45 jack 13
roaming 15, 23, 27, 29
router 300
routing 58, 149
routing domain 62
Routing Information Protocol 153, 170
routing software 104

RS-232 cables 98
RS-422 cable 98
rshd 71
RTS/CTS 242

S

SAP 94, 112, 165
SCP 171
SDU 174
SEC 20
segment 174
Select Network Component Type 209
Select Network Protocol 209
selectable error correction 20
serial connection 98
serial port 98
server, single-port 99
Service Access Point 94, 112
service data unit 174
service provider 165, 264
service user 165
Session Control Protocol 171
session layer 56, 171
Shared Wireless Access Protocol 267
shielded twisted pair 120
Shielded Twisted-Pair Cable 132
short message service 229
Simple Control Protocol 268
Simple Mail Transfer Protocol 172, 300
Simple Network Messaging Protocol 99
single-frequency radio 143
single-port server 99
Small Business Server 275
Small Business Server Internet Connection Wizard 275, 297
smart hub 91
SMS 229
SMTP 172, 279, 300
SNA 87, 90
SNMP 99, 100, 207
socket 62
Spread-spectrum radio transmission 143
spread-spectrum technology 7
spread-spectrum transmission 143
stand-alone wireless network 35
star wiring 40
static IP addressing 58
static mapping 218
STP 120
STP cable 133
structured wiring 38
subnet 106, 107
subnet address 110
subnet mask 59, 60
subnet mask value 60
subnet seed 110
subnetwork 60
switch 100
symbol rate 240
System Network Architecture 87

T

T-connector 128, 129
Tagged Image File Format 172
TCP 115, 300

TCP/IP 47, 48, 50, 52, 54, 55, 61, 74, 75, 82, 148, 149, 199, 236, 276, 300
TCP/IP application 172
TCP/IP connection 64
TCP/IP routing 62
TDMA 17, 21
telnet 152, 172, 197
telnet server 68
terminator 128, 129
TFTP 68
Thicknet 85, 89, 90, 126, 127, 129
Thinnet 85, 126, 127, 129, 151
TIFF 172
time division multiple access 17
Token Ring 76, 87, 94, 236
topology, network 169
topology, ring 169
transceiver cable 130
Transmission Control Protocol 115, 300
Transmission Control Protocol/Internet Protocol 47, 300
transmission media 120, 121
transport layer 56, 170
Trivial File Transfer Protocol 68
TTL 113
tunneling 99
twisted-pair cable 125, 131

U

UDP 62, 68, 113, 115
UMAX 6
Underwriters Laboratories 39
Uniform Resource Locator 301
universally administered address 92
UNIX 49, 63, 65, 151, 207
unshielded twisted pair 4, 120
Unshielded Twisted-Pair Cable 133
upper layer 162
URL 301
USB 8, 13
User Datagram Protocol 62, 115
UTP 4, 38, 120
UTP cable 39, 137, 151
UTP wiring 38

V

vampire tap 130
virtual circuit management 170
Virtual Private Networking 298, 301
Virtual Terminal Protocol 172
Voiceover IP 259
VoIP 259
VPN 301
VTP 172

W

WAN 142, 155, 168
Web browser 196
Web pad 260
Web Publishing Wizard 280, 296
WECA 16, 18, 248
Wide Area Network 142
Wideband Code Division Multiple Access 267
Win32 63

Windows Internet Name Service 211
Windows Internet Naming Service 48, 226
WINS 48, 81, 199, 211, 226
WINS Server Manager 214, 221
WinSock Proxy 295, 299
wired bridging 24
wireless bridge 143
wireless bridging 24
Wireless Equivalent Privacy encryption 14
Wireless Ethernet Compatibility Alliance 16, 248
wireless LAN 16, 20, 22, 24, 30
wireless peer-to-peer network 27
wireless station 237
wiring, daisy-chain 40
wiring, loop 40
wiring, quad 38
wiring, structured 38
wiring, UTP 38
WLAN 19, 235, 240, 247
WLAN Service Area ID 246
World Wide Web 116

X

XNS 93

Z

ZIP 171
Zone Information Protocol 171

EXPLORING MICROSOFT OFFICE XP

Authors: JOHN BREEDEN & MICHAEL CHEEK
ISBN: 079061233X ● **SAMS#:** 61233
Pages: 336 ● **Category:** Computer Technology
Case qty: TBD ● **Binding:** Paperback
Price: $29.95 US/$47.95CAN
About the book: Breeden and Cheek provide an insight into the newest product from Microsoft — Office XP. Office XP is the replacement for Microsoft Office, designed to take users into the 21st century. Breeden and Cheek provide tips and tricks for the experienced office user, to help them find maximum value in this new software.

COMPUTER NETWORKS FOR THE SMALL BUSINESS & HOME OFFICE

Author: JOHN A. ROSS
ISBN: 0790612216 ● **SAMS#:** 61221
Pages: 304 ● **Category:** Computer Technology
Binding: Paperback ● **Price:** $39.95 US/$63.95CAN
About the book: Small businesses, home offices, and satellite offices with unique networks of 2 or more PCs can be a challenge for any technician. This book provides information so that technicians can install, maintain and service computer networks typically used in a small business setting. Schematics, graphics and photographs will aid the "everyday" text in outlining how computer network technology operates, the differences between various network solutions, hardware applications, and more.

To order today or locate your nearest PROMPT® Publications distributor at 1-800-428-7267 or www.samswebsite.com

Prices subject to change.

ADMINISTRATOR'S GUIDE TO SERVERS

Author: LOUIS COLUMBUS
ISBN: 0790612305 ● **SAMS#:** 61230
Pages: 304 ● **Category:** Computer Technology
Case qty: TBD ● **Binding:** Paperback
Price: $39.95 US/$63.95CAN

About the book: Part of Sams Connectivity Series, *Administrator's Guide to Servers* piggybacks on the success of Columbus' best-selling title *Administrator's Guide to E-Commerce*. Columbus takes a global approach to servers while providing the detail needed to utilize the correct application for your Internet setting.

PROMPT® Pointers: Compares approaches to server development. Discusses administration and management. Balance of hands-on guidance and technical information.

Related Titles: *Administrator's Guide to E-Commerce*, by Louis Columbus, ISBN 0790611872. *Exploring LANs for the Small Business and Home Office*, by Louis Columbus, ISBN 0790612291. *Computer Networking for the Small Business and Home Office*, by John Ross, ISBN 0790612216.

Author Information: Louis Columbus has over 15 years of experience working for computer related companies. He has published 10 books related to computers and has published numerous articles in magazines such as *Desktop Engineering, Selling NT Solutions*, and *Windows NT Solutions*. Louis resides in Orange, Calif.

To order today or locate your nearest PROMPT® Publications distributor at 1-800-428-7267 or www.samswebsite.com

Prices subject to change.

ADMINISTRATOR'S GUIDE TO DATAWAREHOUSING

Author: AMITESH SINHA
ISBN: 0790612496 ● **SAMS#:** 61249
Pages: 304 ● **Category:** Computer Technology
Case qty: TBD ● **Binding:** Paperback
Price: $39.95 US/$63.95CAN

About the book: Datawarehousing is the manipulation of the data collected by your business. This manipulation of data provides your company with the information it needs in a timely manner, in the form it desires. This complex and emerging technology is fully addressed in this book. Author Amitesh Sinha explains datawarehousing in full detail, covering everything from set-up to operation to the definition of terms.

PROMPT® Pointers: Covers On-Line Analytical Processing issues. Addresses set-up of datawarehousing systems. Is designed for the experienced IT administrator.

Related Titles: *Designing Serial SANS*, ISBN 0790612461, *How the PC Hardware Works*, ISBN 079061250X.

Author Information: Amitesh Sinha has a Masters in Business Administration and over 10 years of experience in the field of Information Technology. Sinha is currently the Director of Projects with GlobalCynex Inc. based in Virginia and has written numerous articles for computer publications.

To order today or locate your nearest PROMPT® Publications distributor at 1-800-428-7267 or www.samswebsite.com

Prices subject to change.

GUIDE TO CABLING AND COMMUNICATION WIRING

Author: LOUIS COLUMBUS
ISBN: 0790612038 • **SAMS#:** 61203
Pages: 320 • **Category:** Communications
Case qty: TBD • **Binding:** Paperback
Price: $39.95 US/$63.95CAN

About the book: Part of Sams Connectivity Series, *Guide to Cabing and Communication Wiring* takes the reader through all the necessary information for wiring networks and offices for optimal performance. Columbus goes into LANs (Local Area Networks), WANs (Wide Area Networks), wiring standards and planning and design issues to make this an irreplaceble text.

PROMPT® Pointers:
Features planning and design discussion for network and telecommunications applications. Explores data transmission media. Covers Packet Framed-based data transmission.

Related Titles: *Administrator's Guide to E-Commerce*, by Louis Columbus, ISBN 0790611872. *Exploring LANs for the Small Business and Home Office*, by Louis Columbus, ISBN 0790612291. *Computer Networking for the Small Business and Home Office*, by John Ross, ISBN 0790612216.

Author Information: Louis Columbus has over 15 years of experience working for computer-related companies. He has published 10 books related to computers and has published numerous articles in magazines such as *Desktop Engineering, Selling NT Solutions*, and *Windows NT Solutions*. Louis resides in Orange, Calif.

To order today or locate your nearest PROMPT® Publications distributor at 1-800-428-7267 or www.samswebsite.com

Prices subject to change.